COPY 02

INTEGRATED-CIRCUIT OPERATIONAL AMPLIFIERS

INTEGRATED-CIRCUIT OPERATIONAL AMPLIFIERS

Second Edition

George B. Rutkowski, P.E.

Electronic Technology Institute

PRENTICE-HALL, INC., ENGLEWOOD CLIFFS, NEW JERSEY 07632

Library of Congress Cataloging in Publication Data

Rutkowski, George B. (date)
  Integrated-circuit operational amplifiers.

  Previous ed. published as: Handbook of integrated-
circuit operational amplifiers.
  Includes index.
  1. Operational amplifiers. 2. Integrated circuits.
I. Rutkowski, George B. Handbook of integrated-circuit
operational amplifers. II. Title.
TK7871.58.O6R87 1984   621-381'735   83-13773
ISBN 0-13-469007-9

Editorial/production supervision
  and interior design: Barbara H. Palumbo
Cover design: Edsal Enterprises
Manufacturing buyer: Gordon Osbourne

Printed in the United States of America

10 9 8 7 6 5 4 3 2 1

ISBN 0-13-469007-9

PRENTICE-HALL INTERNATIONAL, INC., *London*
PRENTICE-HALL OF AUSTRALIA PTY. LIMITED, *Sydney*
EDITORA PRENTICE-HALL DO BRASIL, LTDA., *Rio de Janeiro*
PRENTICE-HALL CANADA INC., *Toronto*
PRENTICE-HALL OF INDIA PRIVATE LIMITED, *New Delhi*
PRENTICE-HALL OF JAPAN, INC., *Tokyo*
PRENTICE-HALL OF SOUTHEAST ASIA PTE. LTD., *Singapore*
WHITEHALL BOOKS LIMITED, *Wellington, New Zealand*

CONTENTS

ANSWERS TO SELECTED ODD-NUMBERED PROBLEMS   308

INDEX   312

PREFACE

From its inception, the electronics industry has evolved toward smaller, more reliable, and lower cost devices and systems. This evolution has been most rapid in recent years. Some devices, such as integrated-circuit (IC) operational amplifiers (Op Amps), have reached an evolutionary plateau and can be expected to change moderately in the future. Their extreme flexibility, as well as their reliability and economy, have guaranteed them a niche in the world of electronics for many years to come. Metaphorically, we can say that there is an island of stability in the turbulent sea of electronics. It is the isle of the IC Op Amp. There the winds of change blow more slowly and allow the student to plant a firm foothold on the path of a hi-tech electronics career.

Microprocessors (μPs) and microcomputers (μCs) are numerous and popular because of the variety of their potential applications. Most μP and μC applications have not been invented yet but are certain to include interface circuits and systems. An interface enables a μP or μC, which is a digital device or system, to communicate with the outside analog world—the practical business and entertainment world in which we live. IC Op Amps now abound in interface circuits and will continue to abound in systems of the future.

As in the first edition, this edition was developed to:

1. provide an understandable and sufficiently detailed explanation of IC Op Amps and their parameters for practicing or student engineers and technicians;

2. serve as a text with plentiful examples and end-of-chapter problems; and

3. be a reference by containing a collection of Op Amp circuits with means of selecting component values and provide manufacturers' data sheets with tables of IC Op Amp types with their comparative characteristics.

In this edition, more emphasis is placed on the newer Op Amp types and the more common applications. Three-terminal regulators have been added. Pin compatable replacements for the 741, with improved Specs, and packaged instrumentation amplifiers are also included. The discussion of active filters has been expanded to an entire chapter. Programs in BASIC are included in the appendix and can be used to check filter designs.

Chapter 1 of this book contains a review of junction (bipolar) and field-effect (unipolar) transistor fundamentals, principles of differential amplifiers, and the basic inner construction of IC Op Amps. This chapter can be omitted if a *functional block* or *black box* approach is preferred.

IC Op Amps have a peculiar language of their own. Chapter 2 provides definitions and explanations of the essential Op Amp characteristics and compares their practical values to hypothetical ideal ones.

Chapters 3 through 7 provide detailed discussions of the significance of IC Op Amp parameters in practical circuits. Manufacturers' Spec sheets are provided and frequently referenced in a way that engineers and technicians are required to do in practice.

Chapters 8 through 11 show and explain a variety of practical circuit applications. Methods of selecting their component values is especially emphasized.

Chapter 12 discusses fundamentals and applications of current differencing amplifiers and comparator ICs. They offer considerable cost and space savings in some applications.

In addition to manufacturers' Specs, the Appendices contain derivations and many circuit applications of Op Amps and current-differencing amplifiers.

GEORGE B. RUTKOWSKI

INTEGRATED-CIRCUIT OPERATIONAL AMPLIFIERS

1

DIFFERENTIAL
AND OPERATIONAL AMPLIFIERS

The differential amplifier, as its name implies, amplifies the difference between two input signals. These input signals are applied to two separate input terminals, and their difference is called the *differential input voltage* V_{id}. The differential amplifier has two output terminals, and output signals can be taken from either output with respect to ground or across the two terminals themselves. The signals across the output terminals are usually highly amplified versions of the differential input voltages. Differential amplifiers, and variations of them, are found in many applications: electronic voltmeters, instrumentation, industrial controls, signal generators, computers, signal and dc amplifiers, to name only a few.

In this chapter we will first consider the construction and operating fundamentals of junction transistor and field effect transistor differential amplifiers. Then we will see how differential amplifiers can be cascaded, as they are in typical *integrated circuits* (ICs), to obtain very high gain. Finally, we will see how level-shifting circuitry is added to modify the two output terminals of a differential-type circuit to a single output from which a signal can be taken with respect to ground or a common point. Level shifting changes a differential amplifier into an *operational amplifier* (Op Amp).

1.1 TRANSISTOR REVIEW

Since junction transistors and field effect transistors (FETs) are frequently used in discrete and IC differential and operational amplifiers, a review of these active devices is helpful if we expect to obtain even a general understanding of the inner workings of linear ICs.

The *junction transistor* is a current-operated device with three terminals: an emitter E, a collector C, and a base B. In most applications, the relatively small base current I_B is controlled or varied by a signal source. The varying base current I_B in turn controls or varies a much larger collector current I_C and emitter current I_E.

Properly biased transistors are shown in Fig. 1-1. The term *bias* refers to the use of proper dc voltages on the transistor that are necessary to make it work. As shown in Fig. 1-1, the collector C is biased positively with respect to the emitter E on NPN transistors, whereas the collector C is normally negative with respect to the emitter E on PNP transistors. In either case, a NPN or PNP circuit, the base-emitter junction of N- and P-type semiconductors is forward biased. This means that V_{EE} is applied with a polarity that will admit current I_B across the base-emitter junction. The values of V_{EE}, V_S, R_E, and R_B dictate the value of I_B.

As base current flows, it causes a collector current flow, I_C, that is about β* or h_{FE} times larger. That is,

$$I_C \cong \beta I_B \tag{1-1}$$

or

$$I_C \cong h_{FE} I_B \tag{1-2}$$

where β or h_{FE} is specified by the transistor manufacturer. In modern transistors, h_{FE} typically ranges from about 40 to 200. Thus, the base current I_B is typically smaller than the collector current I_C by a factor between 40 and 200.

Note in Fig. 1-1 that in NPN transistor circuits, currents I_B and I_C flow into the transistor while I_E flows out. On the other hand, in PNP circuits, I_B and I_C flow out of the transistor while I_E flows in. By Kirchhoff's current law then

$$I_E = I_C + I_B \tag{1-3}$$

Since I_C is typically much larger than I_B, the base current I_B is often assumed negligible compared to either I_C or I_E. Therefore, Eq. (1-3) can be simplified to

$$I_E \cong I_C \tag{1-4}$$

---

* A transistor's beta (β) is about equal to its h_{FE}. In the symbol h_{FE}, the h means *hybrid* parameter, F means *forward* current transfer ratio, and E means that the *emitter* is common. The capital letters in subscript in h_{FE} specify that this is a dc parameter or dc beta.

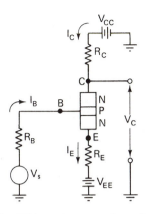

(a) The NPN transistor contains a P-type semiconductor between two N-type materials.

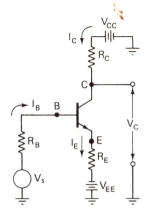

(b) Arrow on emitter points out on NPN transistor symbol.

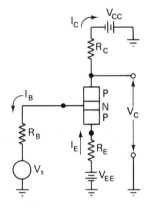

(c) The PNP transistor contains N-type semiconductor between two P-type materials.

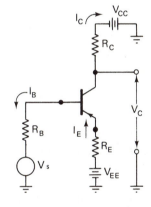

(d) Arrow on the emitter points in on PNP transistor symbol.

Figure 1-1   Properly biased transistor circuits.

Variations of the input voltage V_s on any of the circuits of Fig. 1-1 will cause variations of I_B, which in turn varies I_C flowing through R_C. Thus the voltage across R_C and the output voltage V_C will vary too. Usually, the output variations of V_C are much larger than the input variations of V_s, which means that these circuits are capable of voltage gain—also called voltage amplification A_e.

The *field effect transistor*, FET, unlike the junction transistor, draws negligible current from the signal source V_s and therefore is referrred to as a voltage-operated device. It has three terminals: a source S, a drain D, and a gate G. Properly biased FETs are shown in Fig. 1-2. Basically, variations of a voltage across the gate G and the source S cause drain current I_D and

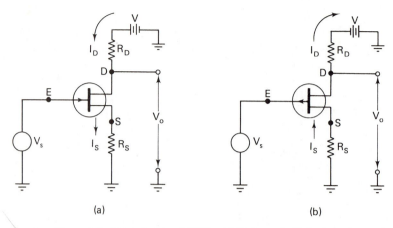

(a)   (b)

Figure 1-2   Properly biased FETs: (a) N-channel, (b) P-channel.

source current I_S variations. Thus, a varying input signal V_s causes a varying gate-to-source voltage V_{GS} which in turn varies I_D, the drop across R_D, and the output voltage V_o. The ratio of the change in drain current I_D to the change in gate-to-source voltage V_{GS} is the FET's transadmittance* y_{fs}, that is,

$$y_{fs} \cong \frac{\Delta I_D}{\Delta V_{GS}} \qquad (1\text{-}5)$$

Because of the very high gate input resistance of the FET, negligible gate current flows, and therefore the drain and source currents are essentially equal, that is, $I_D = I_S$. FET amplifiers are capable of voltage gain, but not as much as are junction transistors. FETs are used where high input resistance is important, such as in some applications of differential and

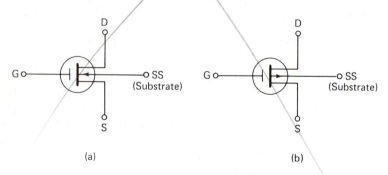

(a)   (b)

Figure 1-3   Symbols used to represent MOSFETs or IGFETs: (a) N-channel, (b) P-channel.

* The forward transadmittance, sometimes called *transconductance*, is referred to by several symbols: y_{fs}, g_{fs}, g_m, g_{21}, etc.

operational amplifiers. As we will see, some types of Op Amps on ICs use FETs in their first stage.

FETs are made in two general types: (1) Junction Field Effect Transistors, JFETs, whose input resistances are on the order of 10^6 to 10^9 ohms, and whose symbols appear in Fig. 1-2, and (2) Metal-Oxide Silicon Field Effect Transistors, MOSFETs, which are also known as Insulated Gate Field Effect Transistors, IGFETs. MOSFETs or IGFETs have even higher gate input resistance values than do JFETs, typically 10^{10} to 10^{14} ohms. Some common symbols in use for MOSFETs are shown in Fig. 1-3.

1.2 THE DIFFERENTIAL CIRCUIT

Figure 1-4a shows a simple junction transistor differential amplifier, and Fig. 1-4b, its typical symbol. If signal voltages are applied to input terminals 1 and 2, their difference V_{id} is amplified and appears as V_{od} across output terminals 3 and 4. Ideally, if both inputs are at the same potential with respect to ground or a common point, causing an input differential voltage $V_{id} = 0$ V, the differential output voltage $V_{od} = 0$ V too, regardless of the circuit's gain.

With input signal sources V_1 and V_2 applied as shown in Fig. 1-4, each

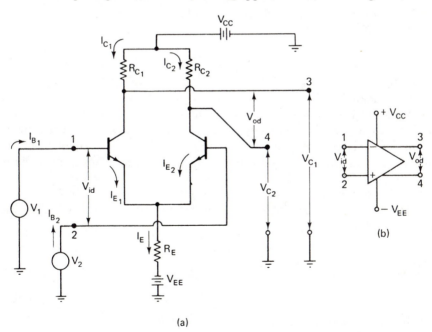

(a)

Figure 1-4   (a) Simple differential amplifier; (b) symbol for the differential amplifier.

base has a dc path to ground, and base currents I_{B_1} and I_{B_2} flow. If the average voltages and internal resistances of the signal sources V_1 and V_2 are equal, then equal base currents flow. This causes equal collector currents, $I_{C_1} = I_{C_2}$, and equal emitter currents, I_{E_1} and I_{E_2}, assuming that the transistors have identical characteristics. Since β or h_{FE} of a typical transistor is much larger than 1, the base current is very small compared to the collector or the emitter current, and therefore the collector and emitter currents are about equal to each other [see Eqs. (1-3) and (1-4)]. For example, in the circuit of Fig. 1-4, $I_{C_1} \cong I_{E_1}$ and likewise $I_{C_2} \cong I_{E_2}$. In this circuit, the current I_E in resistor R_E is equal to the sum of currents I_{E_1} and I_{E_2}. This current can be approximated with the equation

$$I_E = \frac{V_{EE} - V_{BE}}{R_E + R_B/2h_{FE}} \tag{1-6}$$

where V_{BE} is the dc drop across each forward-biased base-emitter junction, which is about 0.7 V in silicon transistors and about 0.3 V in germanium transistors, and

R_B is the dc resistance seen looking to the left of either input terminal, 1 or 2. In the circuit of Fig. 1-4, R_B is the internal resistance of each signal source.

Since the dc source voltage V_{EE} is usually much larger than the base-emitter drop V_{BE} and since R_E is frequently much larger than $R_B/2h_{FE}$, Eq. (1-6) can be simplified to

$$I_E \cong \frac{V_{EE}}{R_E} \tag{1-7}$$

The significance of Eq. (1-7) is that the value of I_E is determined mainly by the values of V_{EE} and R_E and that if V_{EE} and R_E are fixed values, the current I_E is practically constant. Ideally, I_E should be very constant for reasons to be discussed later. The source V_{EE} and resistor R_E, or their equivalents, are often replaced with the symbol for a constant-current source as shown in Fig. 1-5.

The dc voltage to ground at each output terminal is simply the V_{CC} voltage minus the drop across the appropriate collector resistor. Thus the voltage at output 3 to ground is

$$V_{C_1} = V_{CC} - R_{C_1}I_{C_1} \tag{1-8a}$$

Similarly, the voltage at output 4 to ground is

$$V_{C_2} = V_{CC} - R_{C_2}I_{C_2} \tag{1-8b}$$

Their difference is the output differential voltage

$$V_{od} = V_{C_1} - V_{C_2} \tag{1-9}$$

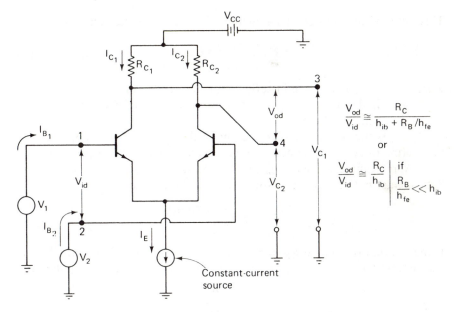

Figure 1-5 Constant-current source driving emitters on differential amplifier.

just as the input differential voltage is

$$V_{id} = V_1 - V_2 \qquad (1\text{-}10)$$

With I_E constant, as in the circuit of Fig. 1-5, the sum of the collector currents, $I_{C_1} + I_{C_2}$, and the sum of the emitter currents, $I_{E_1} + I_{E_2}$, are also constant. Thus, if I_{C_1} is increased, I_{C_2} is forced to decrease, and vice versa. In other words, if the input voltage V_1 is made more positive, base current I_{B_1} increases, increasing I_{C_1} and the voltage drop across R_{C_1}. This in turn causes the voltage V_{C_1} at output 3 to ground to decrease. Furthermore, the increase in I_{C_1} forces I_{C_2} to decrease, which decreases the drop across R_{C_2} and increases the voltage V_{C_2} at output 4. We can therefore reason that if V_1 becomes more positive than V_2, causing a larger differential input voltage V_{id}, then output 3 becomes more negative while output 4 becomes more positive, resulting in a larger differential output voltage V_{od}.

The ratio of the output voltage V_{od} to the input V_{id} is the differential voltage gain A_d of the differential amplifier. Therefore, combining Eqs. (1-9) and (1-10), we can show that

$$A_d = \frac{V_{od}}{V_{id}} = \frac{V_{C_1} - V_{C_2}}{V_1 - V_2} \qquad (1\text{-}11)$$

This differential gain can be estimated with the following equation:

$$A_d \cong \frac{R_C}{h_{ib} + R_B/h_{fe}} \tag{1-12a}$$

where R_C is the resistance in series with each collector,

h_{ib}* is the dynamic emitter-to-base resistance of each transistor,

h_{fe} is the ac β, and

R_B is the resistance seen looking to the left of either input to ground (the internal resistance of either input signal source V_1 or V_2).

Approximate values of h_{ib} are sometimes provided on manufacturers' data sheets, but can also be estimated with the equation

$$\frac{25 \text{ mV}}{I_C} \le h_{ib} \le \frac{50 \text{ mV}}{I_C} \tag{1-13}$$

If the internal resistance R_B of each signal source is small compared to the transistors' h_{ib}, and if the h_{fe}'s are large, Eq. (1-12a) can be simplified to

$$A_d \cong \frac{R_C}{h_{ib}} \tag{1-12b}$$

Example 1-1

Referring to the circuit in Fig. 1-6 and assuming that both transistors are identical having a $\beta = 100$:

(a) Determine the approximate dc voltage to ground at each output and the output differential voltage when both inputs are 0 V (grounded).

(b) Find the approximate differential output voltage at some instant when input $V_1 = 2\text{mV}$ dc to ground and $V_2 = -1\text{mV}$ dc to ground. The internal resistances of the signal sources are negligible. Assume that for each transistor $V_{BE} \ll V_{EE}$.

Answer   (a) We can first estimate the current through R_E using Eq. (1-7):

$$I_E \cong \frac{12 \text{ V}}{6 \text{ k}\Omega} = 2 \text{ mA}$$

Since both inputs are at the same potential, both base currents are equal, causing equal collector currents. Because the sum of the collector currents must be about equal to I_E, each collector has about 1 mA in this case. The voltage V_{C_1} at output 3 is

$$V_{C_1} \cong 12 \text{ V} - 5 \text{ k}\Omega(1 \text{ mA}) = 7\text{V} \tag{1-8a}$$

Similarly, $V_{C_2} = 7$ V as determined with Eq. (1-8b). Therefore, the output differential voltage is

$$V_{od} = 7 \text{ V} - 7 \text{ V} = 0 \text{ V} \tag{1-9}$$

---

* The lowercase letters in subscript in h_{ib} an h_{fe} specify that these are ac parameters and describe the dynamic characteristics of a transistor.

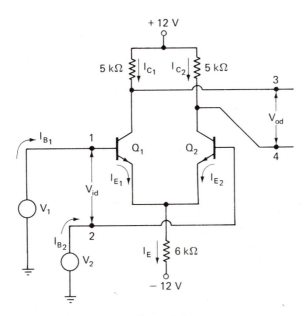

Figure 1-6

(b) With different input voltages, a differential input exists. In this case

$$V_{id} = V_1 - V_2 = 2 \text{ mV} - (-1\text{mV}) = 3 \text{ mV} \qquad (1\text{-}10)$$

This V_{id} is amplified by the circuit's differential gain A_d which can be estimated with the Eq. (1-12b). First we find the range of the dynamic emitter-to-base resistance—that is, the minimum

$$h_{ib} \cong \frac{25 \text{ mV}}{1 \text{ mA}} = 25 \text{ }\Omega$$

and the maximum

$$h_{ib} \cong \frac{50 \text{ mV}}{1 \text{ mA}} = 50 \text{ }\Omega$$

according to Eq. (1-13). Therefore, the maximum differential gain is

$$A_d \cong \frac{5000 \text{ }\Omega}{25 \text{ }\Omega} = 200$$

and the minimum gain is

$$A_d \cong \frac{5000 \text{ }\Omega}{50 \text{ }\Omega} = 100 \qquad (1\text{-}12b)$$

Thus, the differential output voltage can be as large as

$$V_{od} \cong 200(3 \text{ mV}) = 600 \text{ mV}$$

or as low as

$$V_{od} \cong 100(3 \text{ mV}) = 300 \text{ mV}$$

according to Eq. (1-11). The output differential voltage at terminal 3 is negative with respect to terminal 4—i.e., with input 1 more positive than input 2, collector current I_{C_1} is larger than I_{C_2}. This causes a larger drop across R_{C_1} than across R_{C_2}, resulting in less voltage at terminal 3 than at terminal 4 with respect to ground.

A typical FET differential amplifier is shown in Fig. 1-7. It works in much the same way the transistor version does but has much larger resistance as seen looking into its input terminals 1 and 2. Integrated circuit differential and operational amplifiers are available with an FET input stage for applications where extremely large input resistance is needed or where significant dc input bias currents are undesirable. The differential voltage gain of the FET stage can be estimated with the equation

$$A_d = \frac{V_{od}}{V_{id}} \cong \frac{y_{fs}R_D}{R_D y_{os} + 1} \tag{1-14a}$$

where y_{fs} is the forward transadmittance,

$\quad\quad y_{os}$ is the output admittance, and

$\quad\quad R_D$ is the resistance in series with the drain of either FET.

Frequently though, the product $R_D y_{os} \ll 1$, and therefore the Eq. (1-14a), can be simplified to

$$A_d \cong y_{fs}R_D \tag{1-14b}$$

FET parameters y_{fs} and y_{os} are provided on manufacturers' data sheets.

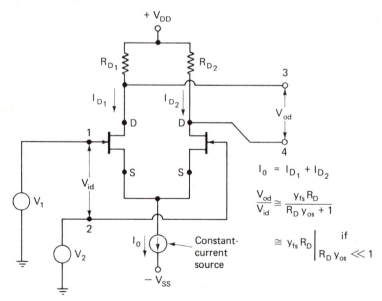

Figure 1-7   FET differential amplifier.

1.3 CURRENT-SOURCE BIASED DIFFERENTIAL AMPLIFIER

As mentioned previously, the emitters of a differential amplifier should ideally be driven by a constant-current source. A constant I_E gives the differential stage excellent *common-mode rejection (CMR)* capability. A differential amplifier has good *CMR* if it does not pass a *common-mode (CM)* voltage. A *CM* voltage is simply an input voltage that appears at both inputs simultaneously. For example, suppose that *both* inputs, 1 and 2, in the circuit of Fig. 1-5, are increased from 0 V to 10 mV above ground. The 10 mV, then, is a common-mode voltage. In this case, it makes both bases more positive, and both transistors attempt to turn on harder. That is, the collector currents of both transistors will attempt to increase, but since their sum is about equal to I_E (a constant), neither of the collector currents increase and the voltages to ground at outputs 3 and 4 remain essentially constant. Thus, if a common-mode voltage is applied, it is not amplified; only differential voltages V_{id} are. In practice, no differential stage has ideal (infinite) *CMR*, and some *CM* voltage appears across the output terminals if a *CM* voltage is applied to its inputs. Typically though, the output common-mode voltage V_{cmo} is much smaller than the input common-mode voltage V_{cmi}. One reason the practical *CMR* is not ideal is that the current I_E driving the emitters is not perfectly constant. Generally, the more constant I_E is, the better the *CMR*.

In practice, the constancy of I_E is improved by driving the emitters with a transistor stage such as in the circuit of Fig. 1-8. The current through transistor Q_3 is very constant even if a varying V_{cmi} is applied or if a change in temperature occurs. The base of Q_3 in biased with the voltage divider containing components R_1, D_1, D_2, and R_2. The diodes D_1 and D_2 help to hold I_E constant even though the temperature changes. Note that current I_1 flows to the node at the base of Q_3 and then divides into paths I_2 and I_{B_3}. If the temperature of Q_3 increases, its base-emitter voltage V_{BE} decreases about 2.5 mV/°C. This reduced V_{BE} tends to raise the voltage drop across R_E and the current I_E. However, the voltage drops across D_1 and D_2 likewise decrease, causing a greater portion of I_1 to contribute to I_2, that is, to flow down through D_1 and D_2. This causes I_{B_3} to decrease, which prevents any significant increase in I_E.

1.4 MULTISTAGE DIFFERENTIAL CIRCUITS

Cascading amplifier stages provides an overall (total) gain that is the product of the individual stage gains. In many applications, such as in linear ICs, differential amplifiers are cascaded for large total gain as shown in Fig. 1-9. A point to note is that the outputs of the first stage are *directly coupled* to the

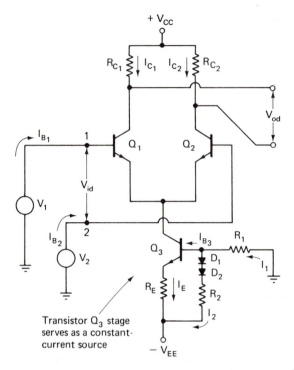

Figure 1-8   Current-source biased differential amplifier.

inputs of the second stage. This presents a problem that requires different biasing methods in these individual stages. For reasons that are discussed in detail in a later chapter, the input (first) stage must be able to take input voltages V_1 and V_2 that might vary positively or negatively with respect to ground, within limits, and have as a result of these input variations, continuous variations of its differential output voltage V_{od}. This means that the transistors or FETs of the input stage should not saturate or cut off with inputs in the neighborhood of about 0 V to ground or common. A transistor is saturated when the current into its base is so large as to reduce its V_{CE} voltage to about zero. It is cut off when its base current is essentially zero, causing the transistor to behave as an open between its collector and emitter. The constant-current source in the input stage holds the emitters in Fig. 1-5 and the sources S in Fig. 1-7 at about ground potential. This allows us to use input voltages that can be varied somewhat around ground potential because the bases must normally be at approximately the same voltage to ground as are the emitters.

     The second stage, however, has a considerable dc component applied to each of its inputs (bases) since the collectors of the first stage are above ground (have a dc voltage to ground) and are applied to the inputs of the second stage. To avoid saturation of the transistors of the second stage, their

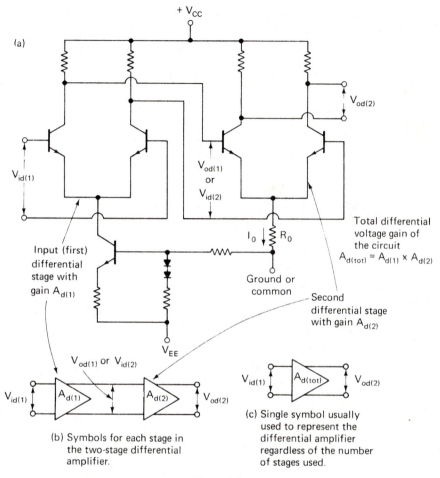

(a)

$V_{id(1)}$

Input (first)
differential
stage with
gain $A_{d(1)}$

$V_{od(1)}$
or
$V_{id(2)}$

I_0 ⬇ R_0

$V_{od(2)}$

Total differential
voltage gain of
the circuit
$A_{d(tot)} = A_{d(1)} \times A_{d(2)}$

Ground or
common

Second
differential stage
with gain $A_{d(2)}$

V_{EE}

$V_{od(1)}$ or $V_{id(2)}$

$V_{id(1)}$ $A_{d(1)}$ $A_{d(2)}$ $V_{od(2)}$

(b) Symbols for each stage in
the two-stage differential
amplifier.

$V_{id(1)}$ $A_{d(tot)}$ $V_{od(2)}$

(c) Single symbol usually
used to represent the
differential amplifier
regardless of the number
of stages used.

Figure 1-9

emitters must be above ground by about the same potential as their bases.
This is accomplished by the use of resistor R_0, one end of which is grounded
instead of terminated to a negative dc bias source. The differential gain of
this two-stage combination is

$$A_{d(\text{tot})} = \frac{V_{od(2)}}{V_{id(1)}} = A_{d(1)} \times A_{d(2)}$$

1.5 LEVEL SHIFTING WITH INTERMEDIATE STAGE

In the previous section we learned that it is often convenient to have both
outputs of a differential amplifier near or at ground potential. Also convenient, for reasons discussed later, is an amplifier with a differential input

(two input leads) but with a single-ended output (one output lead). The output signals can then be taken off this single terminal with respect to ground or a common point. Such a configuration—differential input and single-ended output—is basically an operational amplifier (Op Amp). Op Amps and their applications are thoroughly discussed in this book and are viewed mainly from an external point of view. In this chapter, though, we can look at some internal details and see how an Op Amp is built by adding intermediate and output circuitry to the two-stage differential amplifier.

Since both outputs of the second stage in Fig. 1-9 are normally well above ground, neither of them will serve as an Op Amp output terminal. However, either output can be shifted down to about 0 V to ground with appropriate voltage-divider circuitry. For example, an output of the second differential stage can work into an emitter-follower stage as shown in Fig. 1-10a. Thus a positive voltage at point B can cause 0 V to ground at point E with proper selection of components. That is, if the sum of the voltage drops across the transistor Q and resistor R_a equals the V_{CC} voltage, then the resistor R_b must drop the V_{EE} voltage, and point E is at 0 V with respect to ground. Better results are obtained by using a transistor stage Q_2 as a constant-current source, as shown in Fig. 1-10b, instead of the combination of R_b and V_{EE}. In this case also, point E is near ground potential though point B is well above ground. If point B is driven more positively by the preceding differential stage, transistor Q_1 will drop less voltage and Q_2 will drop more, causing point E to swing positively. Then if point B swings negatively, Q_1's drop increases and Q_2's drop decreases, causing point E to swing negatively.

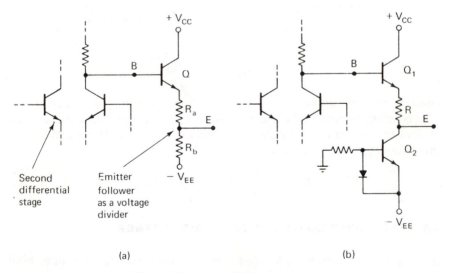

Second differential stage

Emitter follower as a voltage divider

(a)   (b)

Figure 1-10   Level-shifting techniques.

1.6 THE OUTPUT STAGE AND COMPLETE OPERATIONAL AMPLIFIER

Though the output terminal E of the level-shifting stage is not as positive as are the collectors of the preceding differential stage, this output E is not usually used as the output terminal of the Op Amp. An additional output stage, such as in Fig. 1-11, is used. The transistor Q_3 amplifies the signal at point E and drives the *totem-pole* arrangement of transistors Q_4 and Q_5. This output stage not only provides zero or near zero voltage to ground at its output X when the differential input V_{id} to the first stage is zero, but also provides large peak-to-peak output signal capability when an input V_{id} is

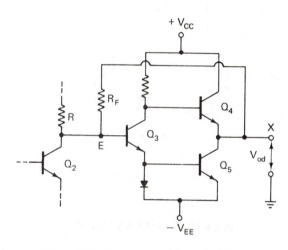

Figure 1-11   Output stage of a typical Op Amp.

applied. In other words, the output X can swing positively to about the V_{CC} voltage and negatively to about the V_{EE} voltage. The resistor R_F provides feedback which stabilizes the output stage, preventing the voltage at X from drifting with temperature changes.

A typical complete operational amplifier is shown in Fig 1-12. By examining this circuit, we can see the stages discussed previously. While specific inner design of IC Op Amps is continually evolving, including FET or Darlington pair inputs, internal compensations, input voltage and output current limiter, and many others, the circuits of this chapter give us a basic understanding of the inner operation of the IC Op Amp. Other components and circuitry contained in some types of Op Amps will be introduced as their relevance becomes apparent.

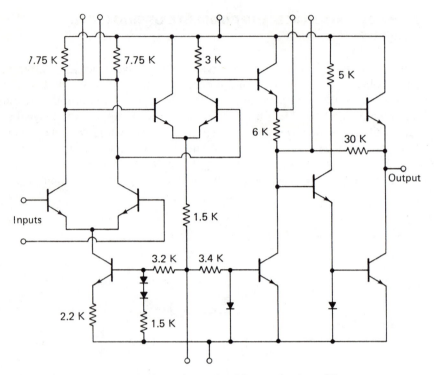

Figure 1-12 Typical complete IC operational amplifier.

REVIEW QUESTIONS

1-1. In which lead of a transistor is the current much smaller than in the other two leads?

1-2. In which leads of a transistor are the currents approximately equal?

1-3. The same current flows in which leads of an FET?

1-4. The drain current in an FET circuit is controlled by a voltage across what leads?

1-5. What does a differential amplifier do?

1-6. The junction transistor differential amplifier has (*higher*) (*lower*) input resistance and a (*higher*) (*lower*) voltage gain than does the FET version.

1-7. What is the purpose of cascading differential amplifiers?

1-8. From the external point of view, how does an Op Amp differ from a differential amplifier?

1-9. What is the purpose of an intermediate stage following the second differential amplifier as shown in Fig. 1-12?

1-10. Generally, over what approximate range is the output voltage of the Op Amp in Fig 1-12 able to swing?

PROBLEMS

1-1. If a transistor's collector current $I_C = 0.5$ mA and its $h_{FE} = 80$, what is its approximate base current I_B?

1-2. If in the circuit of Fig. 1-4 both inputs are grounded, $V_{EE} = V_{CC} = 10$ V, $R_{C_1} = R_{C_2} = 7.2$ kΩ, and $R_E = 10$kΩ, what is the voltage at each collector with respect to ground, and what is the differential output voltage? (Assume that the transistors are identical.)

1-3. Referring to the circuit described in the previous problem what approximate voltage can we expect at each output to ground, and what is the differential output voltage if +400 mV dc is applied to each input?

1-4. Referring to the circuit described in Prob. 1-2, approximately what dc base current flows in each input terminal if the β of each transistor is 200?

1-5. What is the approximate range of voltage gain of the circuit described in Prob. 1-2 assuming that the internal resistances, of V_1 and V_2 are negligible?

1-6. If in the circuit of Fig. 1-7, $I_o = 0.4$ mA, $V_{DD} = 12$ V, and $R_{D_1} = R_{D_2} = 20$ kΩ, what is the approximate drain-to-ground voltage on each FET when both gates are grounded? (Assume that the FETs are identical.)

2

CHARACTERISTICS
OF OP AMPS
AND THEIR POWER
SUPPLY REQUIREMENTS

The term *operational amplifier* refers to a high-gain dc amplifier that has a differential input (two input leads) and a single-ended output (one output lead). The signal output voltage V_o is larger than the differential input signal across the two inputs by the gain factor of the amplifier (see Fig. 2-1). Op Amps have characteristics such as high input resistance, low output resistance, high gain, etc., that make them highly suitable for many applications, a number of which are shown and discussed in later chapters. By examining some applications and comparing the characteristics of typical Op Amps, we will see that some types are apparently better than others. The overall most desirable characteristics that Op Amp manufacturers strive to obtain in their products are *ideal characteristics*. While some of these ideal characteristics are impossible to obtain, we often assume that they exist to develop circuits and equations that work perfectly on paper. These circuits and equations then work very well with practical Op Amps, provided their characteristics are not too far from the ideal values. Actual Op Amp characteristics are more or less ideal, relative to conditions external to the Op Amp: signal source resistance, load resistance, amount of feedback used, etc. Some of the more important characteristics, ideal and practical values, are given and defined in the following sections.

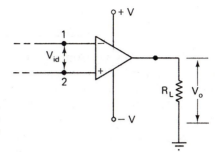

Figure 2-1 Typical Op Amp symbol. $+V$ is the positive dc power supply voltage; $-V$ is the negative dc power supply voltage; $A_{VOL} = V_o/V_{id}$.

2.1 OPEN-LOOP VOLTAGE GAIN* A_{VOL}

The open-loop voltage gain A_{VOL} of an Op Amp is its *differential* gain under conditions where no negative feedback is used as shown in Figs. 2-1 and 2-2. Ideally, its value is infinite—that is, the ideal Op Amp has an open-loop voltage gain

$$A_{VOL} = \frac{V_o}{V_{id}} = -\infty \qquad (2\text{-}1\text{a})$$

or

$$A_{VOL} = \frac{V_o}{V_1 - V_2} = -\infty \qquad (2\text{-}1\text{b})$$

The negative sign means that the output V_o and the input V_{id} are out of phase. The concept of an infinite gain is difficult to visualize and impossible to obtain. The important point to understand is that the Op Amp's output voltage V_o should be very much larger than its differential input V_{id}. To put it another way, the input V_{id} should be infinitesimal compared to any practical value of output V_o. Open-loop gains A_{VOL} range from about 5000 (about 74 dB) to 100,000 (about 100 dB)—that is,

$$5000 \leq A_{VOL} \leq 100{,}000 \qquad (2\text{-}2\text{a})$$

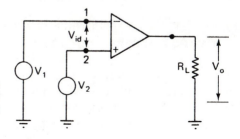

Figure 2-2 Op Amp with input voltages V_1 and V_2 whose difference is V_{id}. Power supply voltages $+V$ and $-V$ are assumed if not shown.

* The open-loop voltage gain is referred to with a variety of symobls: A_{VOL}, A_d, A_{EOL}, A_{VD}, etc.

or

$$74 \text{ dB} \leq A_{VOL} \leq 100 \text{ dB} \tag{2-2b}$$

with popular types of Op Amps. The fact that the output V_o is 5000 to 100,000 times larger than the differential input V_{id} does not mean that V_o can actually be very large. In fact, its positive and negative peaks are limited to values a little less than the positive and negative supply voltages being used to power the Op Amp. Since the dc supply voltages of IC Op Amps are usually less than 20 V, the peaks of the output voltage V_o are less than 20 V. This fact coupled with the high gain factor of the typical Op Amp makes the voltage V_{id} across the inputs 1 and 2 very small. Of course, the larger the open-loop gain A_{VOL}, the smaller V_{id} is in comparison to any practical value of V_o. Thus, since

$$A_{VOL} = \frac{V_o}{V_{id}} \tag{2-3a}$$

or

$$A_{VOL} = \frac{V_o}{V_1 - V_2} \tag{2-3b}$$

then

$$V_{id} = \frac{V_o}{A_{VOL}}$$

and, therefore,*

$$\lim_{A_{VOL} \to \infty} V_{id} = 0$$

Thus, if the gain A_{VOL} in the expression V_o/A_{VOL} is very large, the value of this expression, which is V_{id}, must be relatively very small. In fact, V_{id} is usually so small that we can assume there is practically no potential difference between the inverting and noninverting inputs.

2.2 OUTPUT OFFSET VOLTAGE V_{oo}

The output offset voltage V_{oo} of an Op Amp is its output voltage to ground or common under conditions when its differential input voltage $V_{id} = 0$ V. Ideally $V_{oo} = 0$ V. In practice, due to imbalances and inequalities in the differential amplifiers within the Op Amp itself (see Fig. 1-12), some output

---

* As the value of open-loop gain A_{VOL} approaches infinity, the value of differential input voltage V_{id} approaches zero if $V_o \neq 0$.

offset voltage V_{oo} will usually occur even though the input $V_{id} = 0$ V. In fact, if the open-loop gain A_{VOL} of the Op Amp is high and if no feedback is used, the output offset is large enough to saturate the output. In such cases, the output voltage is either a little less than the positive source voltage $+V$ or the negative source $-V$. This is not as serious as it first appears because corrective measures are fairly simple to apply and because the Op Amp is seldom used without some feedback. When corrective action is taken to bring the output to 0 V when the applied differential input *signal* is 0 V, the Op Amp is said to be *balanced* or *nulled*.

If both inputs are at the same finite potential, causing $V_1 - V_2 = V_{id} = 0$ V, and if the output $V_o = 0$ V as a result, the Op Amp is said to have an ideal *common-mode rejection (CMR)*. Though most practical Op Amps have a good *CMR* capability, they do pass some *common-mode (CM)* voltage to the load R_L. Typically though, the load's common-mode output voltage V_{cmo} is hundreds or even thousands of times *smaller* than the input common-mode voltage V_{cmi}. A thorough discussion of the practical Op Amp's *CMR* capability and its applications is given in a later chapter.

2.3 INPUT RESISTANCE R_i

The Op Amp's input resistance R_i is the resistance seen looking into its inputs 1 and 2 as shown in Fig. 2-3. Ideally, $R_i = \infty$ Ω. Practical Op Amps' input resistances are not infinite but instead range from less than 5 kΩ to over 20 MΩ, depending on type. Though resistances in the low end of this range seem a bit small compared to the desired ideal (∞ Ω), they can be quite large compared to the low internal resistances of some signal sources commonly used to drive the inputs of Op Amps. Generally, if a high-resistance signal source is to drive an Op Amp, the Op Amp's input resistance should be relatively large. As we will see later, the *effective* input resistance R_i' is made considerably larger than the manufacturer's specified R_i by wiring the Op Amp to have negative feedback. Manufacturers usually specify their Op Amps' input resistances as measured under open-loop conditions (no feedback). In most linear applications, Op Amps are wired with some feedback and this improves (increases) the effective resistance seen by the signal source driving the Op Amp.

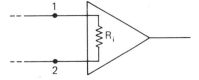

Figure 2-3 Input resistance R_i is the resistance seen looking into the inputs 1 and 2.

2.4 OUTPUT RESISTANCE R_o

With a differential input signal V_{id} applied, the Op Amp behaves like a signal generator as the load connected to the output sees it. As shown in Fig. 2-4, the Op Amp can be shown as a signal source generating an open-circuit voltage of $A_{VOL}V_{id}$ and having an internal resistance of R_o. This R_o is the Op Amp's output resistance and ideally should be 0 Ω. Obviously, if $R_o = 0$ Ω in the circuit of Fig. 2-4, all of the generated output signal $A_{VOL}V_{id}$ appears at the output and across the load R_L. Depending on the type of Op Amp, specified output resistance R_o values range from a few ohms to a few hundred ohms and are usually measured under open-loop (no feedback) conditions. Fortunately, the *effective* output resistance R'_o is reduced considerably when the Op Amp is used with some feedback. In fact, in most applications using feedback, the effective output resistance of the Op Amp is very nearly ideal (0 Ω). The effect feedback has on output resistance is discussed in more specific terms later.

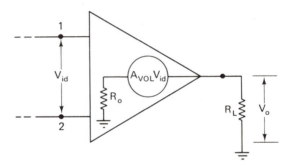

Figure 2-4   Output resistance R_o tends to reduce output voltage V_o.

2.5 BANDWIDTH *BW*

The *bandwidth BW* of an amplifier is defined as the range of frequencies at which the output voltage does not drop more than 0.707 of its maximum value while the input voltage *amplitude* is constant. Ideally, an Op Amp's bandwidth $BW = \infty$. An infinite bandwidth is one that starts with dc and extends to infinite cycles/second (Hz). It is indeed idealistic to expect such a bandwidth from any kind of amplifier. Practical Op Amps fall far short of this ideal. In fact, limited high-frequency response is a shortcoming of Op Amps. While some types of Op Amps can be used to amplify signals up to a few megahertz, they require carefully chosen externally wired compensating components. Most general purpose types of IC Op Amps are limited to less than 1 MHz bandwidth and more often are used with signals well under a few

kilohertz, especially if any significant gain is expected of them. The limited
high-frequency response of Op Amps is not a serious matter with most types
of instrumentation in which Op Amps are used extensively.

2.6 RESPONSE TIME

The response time of an amplifier is the time it takes the output of an
amplifier to change after the input voltage changes. Ideally, response time =
0 seconds, that is, the output voltage should respond instantly to any change
on the input. Figure 2-5 shows an Op Amp's typical output-voltage response
to a step input voltage when wired for unity gain. Manufacturers specify a
slew rate that gives the circuit designer a good idea of how quickly a given
Op Amp responds to changes of input voltage. Note that the time scale in
Fig. 2-5 is in microseconds (μs), which indicates that the typical Op Amp
responds quickly though not instantly. Also note that the output voltage,
when changing, overshoots the level it eventually settles at. Overshoot is the
ratio of the amount of overshoot to the steady-state deviation expressed as a
percentage. For example, if we observe the responding output voltage of
Fig. 2-5 on an oscilloscope whose vertical deflection sensitivity is set at 10
V/cm, and the amount of overshoot measures 0.2 cm, the percentage of
overshoot is

$$\frac{\text{Amount of overshoot}}{\text{Steady-state deflection}} \times 100 = \frac{0.2 \text{ cm}}{2 \text{ cm}} \times 100 = 10\%$$

From time to time we will refer to the various *ideal* characteristics, and
therefore the following summary of them will be helpful:

(1) Open-loop voltage gain $A_{VOL} = \infty$.

(2) Output offset voltage $V_{oo} = 0$ V.

(3) Input resistance $R_i = \infty \ \Omega$.

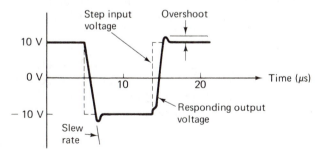

Figure 2-5   Typical Op Amp response to a step input voltage when wired as a
voltage follower (unity gain).

(4) Output resistance $R_o = 0 \ \Omega$.

(5) Bandwidth $BW = \infty$ hertz.

(6) Response time $= 0$ secs.

2.7 POWER SUPPLY REQUIREMENTS

In many applications, the Op Amp's output voltage V_o must be capable of swinging in both positive and negative directions. In such applications, the Op Amp requires two source voltages: one positive $(+V)$ and the other negative $(-V)$ with respect to ground or a common point. These dc source voltages must be well filtered and regulated, otherwise the Op Amp's output voltage will vary with the power supply variations. The output voltage of an Op Amp varies more or less with power supply variations, depending on its closed-loop* voltage gain and sensitivity factor S which is usually specified on the manufacturer's data sheets. An ideal Op Amp has a sensitivity factor $S = 0$, which means that power supply voltage variations have no effect on its output. Practical Op Amps, however, are affected by changes in the supply voltages, and therefore regulated supplies are used to keep Op Amp outputs responsive to differential input voltages only.

2.8 ZENER REGULATORS

Some power supplies use Op Amps as part of their voltage regulating system, and these are discussed in a later chapter. If the current drain is not too large and if it is relatively constant, simple zener-diode-regulated† power supplies such as in Fig. 2-6 can be used. Some types of IC Op Amps can work with dc source voltages up to and over ± 20 V, but usually values under ± 15 V are recommended.‡ Thus, in practice, the output voltages of the power supplies in Fig. 2-6 are ± 15 V or under. The circuits in Fig. 2-6c and d show how the dc power sources are sometimes drawn. Usually these dc sources and the pins to which they are connected are not shown but are assumed to be there (see Figs. 2-2, 2-3, and 2-4).

Generally, the ac voltages e in the circuits of Fig. 2-6a and b are selected so that their peaks exceed the actual required dc outputs $+V$ or $-V$ by 10% to 40%, depending on how unregulated this ac voltage e is. Capacitors C_1 and C_2 charge to approximately the peak of ac input e because of the rectifying action of diodes D_1 and D_2 or of diodes D_a through D_d. If

---

* The closed-loop gain of an Op Amp is its gain when a feedback loop is used.

† Zener diodes are also called *regulator* diodes on typical data sheets.

‡ Some hybrid Op Amps, containing discrete components and ICs, are designed to work with dc supply voltages over 100 V and hundreds of volts peak-to-peak output capability.

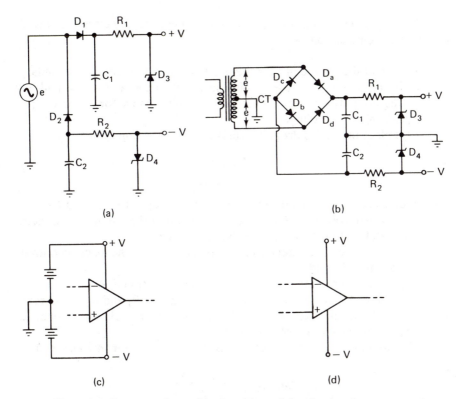

Figure 2-6 Power supply capable of positive and negative dc voltages to ground or a common point: (a) half-wave zener-regulated circuit; (b) full-wave zener-regulated circuit; (c) battery symbols sometimes used to represent plus and minus source voltages; (d) common method of showing dc source voltage.

voltage e is sinusoidal, each capacitor's voltage is about $e \sqrt{2}$ volts, where e is the rms value. The difference between each capacitor's voltage and its respective zener voltage appears across the resistor. Thus, the difference between the voltages across C_1 and D_3 appears across R_1. Similarly, the drop across R_2 is the difference between the voltages across C_2 and D_4. The zeners can normally conduct a varying current and hold a fairly constant voltage drop. This characteristic enables them to act as voltage regulators. Thus, if the ac input voltage e increases or decreases, as an unregulated ac source typically does, each zener draws more or less current, respectively. This increases or decreases the drop across each resistor, but the zener and output voltages remain quite constant. On the other hand, if voltage e is constant but the loads connected to the $+V$ and $-V$ output terminals draw more or less current, the zeners respond by drawing less or more current, respectively, to keep the drop across the resistors and zeners quite constant. The point is, the zeners have reasonably constant voltage drops across them even though the ac input voltage e and/or the load resistances vary.

The component values for the regulated power supply circuits in Fig. 2-6 can be selected as follows:

(1) Select the source of ac voltage e (usually a transformer secondary) so that its peak is about 10% to 40% larger than the actual $+V$ and $-V$ outputs required. The dc voltage across each filter capacitor is about equal to the peak of the ac voltage e.

(2) Select the series dropping resistors with the equation

$$R < \frac{E_{\min} - V_z}{I_{L(\max)}} \tag{2-4}$$

where $E_{\min}$ is the minimum dc across each of the capacitors C_1 and C_2, V_z is the zener voltage of D_3 or D_4 (assuming that the circuit is symmetrical and that both zeners are the same type),

$I_{L(\max)}$ is the maximum current drawn by each of the loads on outputs $+V$ and $-V$, and

R is the approximate value of each resistor R_1 and R_2.

After the value of R is calculated, the actual values of R_1 and R_2 are selected to be slightly less. This allows for some current flow in each of the zeners at all times and prevents loss of regulation.

(3) The power rating of each series dropping resistor must exceed

$$P_{R(\max)} = \frac{(E_{\max} - V_z)^2}{R} \tag{2-5}$$

where $E_{\max}$ is the maximum dc voltage across each capacitor C_1 and C_2.

(4) The power rating of each zener must exceed

$$P_{z(\max)} = \left[\frac{E_{\max} - V_z}{R} - I_{L(\min)} \right] V_z \tag{2-6}$$

where R is the actual value of each of the resistors R_1 and R_2 used, and $I_{L(\min)}$ is the minimum current each load draws from the dc output terminals $+V$ and $-V$.

(5) As a rule of thumb, the maximum secondary rms current rating of the transformer, or the output-current capability of any source of the ac voltage e, should exceed the maximum dc current flow through R_1 and R_2. More specifically, the current capability of the source e must exceed four times the maximum current in R_1 or R_2 in the circuit of Fig. 2-6a. In Fig. 2-6b, the transformer secondary-current (I_s) capability should exceed 1.8 times the maximum current in R_1 or R_2. Thus, for the circuit in Fig. 2-6a,

$$I_s > 4I_{R(\max)} = 4 \left[\frac{E_{\max} - V_z}{R} \right] \tag{2-7a}$$

And for the circuit in Fig. 2-6b,

$$I_s > 1.8I_{R(max)} = 1.8 \left[\frac{E_{max} - V_z}{R} \right] \tag{2-7b}$$

(6) The *PIV* (peak inverse voltage) ratings of the rectifier diodes, D_1 and D_2 or D_a through D_d, must exceed twice the maximum peak of voltage e. Thus, if e is the rms value, then

$$PIV > 2(e\sqrt{2}) \tag{2-8}$$

and generally, their current ratings should exceed the maximum expected secondary rms current as determined with the appropriate version of Eq. (2-7).

(7) At 60 Hz, for reasonable filtering capability, each of the filter capacitors should have a capacitance of at least

$$C \cong 50 \text{ ms} \left(\frac{I_{R(max)}}{V_z} \right) \tag{2-9a}$$

for the half-wave circuit such as in Fig. 2-6a, or at least

$$C \cong 25 \text{ ms} \left(\frac{I_{R(max)}}{V_z} \right) \tag{2-9b}$$

for the full-wave circuit such as in Fig. 2-6b.

Example 2-1

Suppose a transformer is the source of the ac voltage e in the circuit of Fig. 2-6a, and that its secondary is specified to be 12 V rms when the primary is 115 V rms. If the output voltages are to be ±12 V dc, each of the loads on the $+V$ and $-V$ outputs draws 0 to 40 mA, and the primary voltage varies form 110 V to 125 V rms. Determine:

(a) the resistances of R_1 and R_2 and their power ratings,
(b) the zeners' voltages V_z and the required power rating of each,
(c) the transformer's approximate required secondary-current capability,
(d) the rectifier diodes' *PIV* and current ratings, and
(e) the filter capacitors' approximate values.

Answer.   This transformer's turns ratio $N_p/N_s = 115$ V/12 V $\cong 9.6$. Thus, when the primary voltage is 110 V (minimum), the minimum secondary voltage $e_{min} \cong 110$ V/9.6 $\cong 11.5$ V rms, and therefore the minimum dc voltage across each capacitor is

$$E_{min} \cong 11.5\sqrt{2} \cong 16.2 \text{ V}$$

When the primary voltage is 125 V, the maximum secondary voltage $e_{max} \cong 125$ V/9.6 $\cong 13$ V rms, which causes a maximum dc voltage across each capacitor of

$$E_{max} = 13\sqrt{2} \cong 18.4 \text{ V}$$

Since we require ± 12-V outputs, each zener is selected so that $V_z \cong 12$ V; therefore,
(a)

$$R < \frac{16.2 - 12}{40 \text{ mA}} = \frac{4.2 \text{ V}}{40 \text{ mA}} \cong 105 \ \Omega \qquad (2\text{-}4)$$

Since R_1 and R_2 are to be slighly less than this calculated R, we look for available resistors under 105 Ω. Standard 100-Ω resistors will probably be unsuitable, considering their normal tolerances. For example, a resistor rated at 100 Ω with 20% or 10% tolerance can have a resistance of more than 105 Ω. A commonly available, next-size-smaller, standard value is 75 Ω; therefore, let $R_1 = R_2 = 75 \ \Omega$ in this case. This leads to each resistor's maximum power dissipation:

$$P_{R(max)} = \frac{(18.4 - 12)^2}{75 \ \Omega} = \frac{6.4^2}{75 \ \Omega} \cong 0.55 \text{ W} \qquad (2\text{-}5)$$

A 3/4-watt or 1-watt rating will probably work well in most environments.

(b) We've already selected each zener's voltage to be 12 V. Each one's maximum power dissipation is

$$P_{z(max)} \cong \left[\frac{18.4 - 12}{75} - 0 \right] 12 \cong 1 \text{ W} \qquad (2\text{-}6)$$

Zeners with a power rating of 1 watt or more should be used, depending on the ambient temperature. Maximum power dissipation versus temperature curves, or derating factors, are provided by the zener manufacturers.

(c) With the values of E_{max}, V_z, and R known, the maximum current in R may be calculated:

$$I_{R(max)} = \frac{18.4 - 12}{75} = \frac{6.4 \text{ V}}{75 \ \Omega} \cong 85.4 \text{ mA}$$

Therefore, by Eq. (2-7a), the required current capability of the secondary is

$$I_s \cong 4(85.4 \text{ mA}) \cong 342 \text{ mA}$$

Like the transformer's secondary, each rectifier diode should be capable of at least 342 mA in this case.

(d)

$$PIV > 2(13 \text{ V } \sqrt{2}) = 36.8 \text{ V} \qquad (2\text{-}8)$$

(e) Since this circuit is a half-wave type, a filter capacitance

$$C = 50 \text{ ms} \left[\frac{I_{R(max)}}{V_z} \right] = 50 \text{ ms} \left[\frac{85.4 \text{ mA}}{12 \text{ V}} \right] \cong 350 \ \mu\text{F}$$

will keep the ac ripple output negligible for most practical purposes.

2.9 THREE-TERMINAL REGULATORS

Manufacturers of semiconductor components make an impressive variety of inexpensive IC regulators that are easy to use. The simplest—three-terminal regulators—are able to provide fixed regulated output voltages. As shown in Fig. 2-7, three-terminal regulators are connected between the unregulated dc voltage source V_{in} and the load that requires the regulated voltage V_{out}. A partial list of common three-terminal regulators, given in Appendix A, shows some typical regulated output voltages: 6 V, 8 V, 12 V, 15 V, etc. Many types of three-terminal regulators can deliver over 1 A to their load while regulating V_{out}.

In Fig. 2-7a, the LM340K-15 is a positive 15-V regulator.* If the regulator is located more than 5 cm (about 2 inches) from the filter capacitor C, use of an input capacitor C_{in}, which helps to maintain stability, is recommended. If needed, C_{in} should be at least a 0.22 μF ceramic disk, a 2 μF solid tantalum, or a 25 μF electrolytic. The size of the filter capacitor C can be selected with the use of Eq. 2-9b. Simply substitute $I_{R(max)}$ with the maximum current that the load draws from the regulator. An output capacitor C_{out} serves to limit high-frequency noise and improve transient response. Good transient response on a regulator means that its output voltage V_{out} remains constant (regulated) even though there are rapid changes in the load current.

In Fig. 2-7b, the LM320K-15 is a negative 15-V regulator. As with the positive regulator, C_{in} is added if there is some distance between the filter capacitor C and the regulator. The chip (IC) manufacturers recommend that C_{out} of 1 μF or larger solid tantalum, or a 25 μF or larger aluminum electrolytic, be used to assure stability.

As shown in Fig. 2-7c, positive and negative regulators can work together in a dual supply that has both 15 V and −15 V outputs. The diodes D are recommended if this type of dual supply is to provide power to a common load, such as Op Amps.

The unregulated voltage V_{in} typically varies with changes in the ac line voltage and load current. The minimum value of V_{in} must always be larger than the regulated output V_{out} by the *dropout* voltage. Dropout voltages are specified by the regulator manufacturers and are typically about 2 V. Also specified are quiesent currents, I_Q. For a given regulator, the quiesent current is the difference in the input and load currents. It flows in pin 3 of the LM340K-15 and in pin 1 of the LM320K-15 in the circuits of Fig. 2-7.

Example 2-2

If in the circuit of Fig. 2-7a, V_{in} varies from 20 V to 28 V and the load draws zero to 150 mA, determine: (a) the minimum and maximum voltages across pins 1 and 2; (b) the dropout voltage and quiesent current of the regulator (refer to the table of

---

* See Appendix B for an interpretation of the alphanumerically coded numbers.

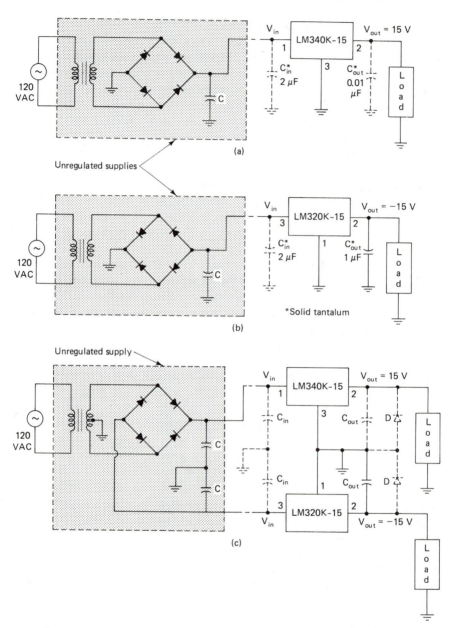

Figure 2-7 Three-terminal regulators: (a) positive regulator, (b) negative regulator, and (c) dual supply.

Appendix A); and (c) the minimum and maximum currents that this regulator draws from the bridge rectifier.

Answers.   (a) According to Appendix A, the output voltage at pin 2 is 15 V. Since the unregulated voltage at pin 1 varies from 20 V to 28 V, the voltage across these pins can be as low as 20 V − 15 V = 5 V or as high as 28 V − 15 V = 13 V.

   (b) Appendix A shows that the LM340K-15 has a specified minimum input voltage of 17 V and an output of 15 V. Their difference, 2 V, is the dropout voltage. If V_{in} becomes less than 17 V, the regulator drops out of regulation.

   (c) When the load draws 0 A, the regulator draws 6 mA: the quiescent current. When the load draws 150 mA, the regulator requires 156 mA.

2.10 REGULATOR POWER DISSIPATION

Three-terminal regulators dissipate power while performing their regulating function. The approximate value of maximum power dissipation $P_{D(max)}$ can be determined by multiplying the maximum voltage across the regulator's input and output terminals by the maximum load current—that is,

$$P_{D(max)} \cong (V_{in(max)} - V_{out})I_{L(max)} \qquad (2\text{-}10)$$

Power dissipation P_D causes heat at the junctions of the transistors within the regulator. This heat flows from the junctions to the case (package) of the regulator. The heat then flows from the case to the surrounding air or to a heat sink. If used, a heat sink, in turn, transfers the heat to the air. Needless to say, $P_{D(max)}$ should never exceed the power rating of the regulator which is sometimes specified.

2.11 HEAT SINKS

Regulators are available in a variety of package types (see appendices B and C). The package of a regulator is called its *case*. Often the case serves as the ground lead (pin) on positive regulators and as the input lead on negative regulators. When required to dissipate about 1 W or more, regulators are housed in packages that are made to be mounted on a heat sink. If the case is not the ground lead, an electrically insulating washer is used between the case and the sink. Heat-conducting silicone grease is often used between the case and sink, or on both sides of the washer.

   Heat generated in a regulator must first flow through its *junction-to-case* thermal resistance θ_{JC}. From the case, the heat flow continues through a *case-to-sink* thermal resistance θ_{CS} (see Table 2-1). Finally, the heat flow is through the *sink-to-air* thermal resistance θ_{SA}.

TABLE 2-1.   Thermal Resistance of Insulating Washers

Material	Dry	With Silicone
Mica	0.8	0.4
Teflon	1.45	0.8
Anodized Aluminum	0.4	0.35
No Insulator	0.2	0.1

The amount of power, in the form of heat, that flows from a regulator's junctions to the surrounding air can be determined by an equation similar to Ohm's law:

$$P_D = \frac{T_J - T_A}{\theta_{JC} + \theta_{CS} + \theta_{SA}} \tag{2-11}$$

where P_D is the dissipated power (heat) flow from the junctions to air in watts.

T_J is the temperature of the transistors' junctions within the regulator in degrees Celsius (°C).

T_A is the ambient (surrounding air) temperature in °C.

θ_{JC} is the junction-to-case thermal resistance, in °C/W, which is specified by the regulator manufacturer.

θ_{CS} is the case-to-sink thermal resistance, in °C/W, which is obtained from Table 2-1 or its equivalent.

θ_{SA} is the sink-to-air thermal resistance, in °C/W, which is specified by the manufacturer of the heat sink.

The term $T_J - T_A$ is the difference in the junction and ambient air temperatures. As larger voltages cause larger currents in electrical circuits, so likewise do larger temperature differences cause larger power (heat) flow. Also, as smaller circuit resistance admits more current, then similarly, smaller thermal resistance allows for increased heat flow. If we are to select the smallest heat sink necessary, we need to know the largest θ_{SA} that will be adequate; physically larger heat sinks have smaller values of θ_{SA}. We can start by rearranging Eq. 2-11 to the following:

$$\theta_{SA} = \frac{T_J - T_A}{P_D} - \theta_{JC} - \theta_{CS} \tag{2-12}$$

When using the above equation, make P_D the maximum power that the regulator is required to dissipate and T_J the maximum recommended temperature which usually is provided on the spec sheets.

Example 2-3

Referring to the circuit of Fig. 2-7a, the transformer has a 15-V secondary when 115 V ac is on its primary (see row 7 of Table 2-2). The line voltage varies from 105 V to 130 V and the load draws from 0 mA to 200 mA. The regulator's $\theta_{JC} = 3$°C/W and working temperature range is -20°C to 150°C. The case will be directly mounted on a

TABLE 2-2.   **Power Transformers, 115 VAC Primary, 50-60 Hz**

Output Watts	Secondary	
	Series Conn.	Parallel Conn.
1½	8 V CT at .188 A	4 V at .376 A
1½	15 V CT at .100 A	7.5 V at .200 A
1½	30 V CT at .050 A	15 V at .100 A
1½	54 V CT at .028 A	27 V at .056 A
1½	76 V CT at .020 A	38 V at .040 A
4½	8 V CT at .562 A	4.0 V at 1.124 A
4½	15 V CT at .300 A	7.5 V at .600 A
4½	30 V CT at .150 A	15 V at .300 A
4½	54 V CT at .084 A	27 V at .168 A
4½	76 V CT at .060 A	38 V at .120 A
7½	8 V CT at .940 A	4 V at 1.880 A
7½	15 V CT at .940 A	7.5 V at 1.000 A
7½	30 V CT at .250 A	15 V at .500 A
7½	54 V CT at .140 A	27 V at .280 A
7½	76 V CT at 1.00 A	38 V at .200 A

heat sink; no insulator or silicone grease will be used (refer to Table 2-1). Find the maximum power dissipated by this regulator and the largest θ_{SA} (smallest heat sink) that we can use. If this regulator's dropout voltage is 2 V, will it reliably regulate over the entire input voltage range? Assume that the ambient temperature is 25°C.

Answers.   This transformer has a voltage ratio of 115 V/15 V $\cong$ 7.67/1. The filter capacitor C will charge to about the peak value of the secondary. Therefore,

$$V_{in(min)} \cong (105 \text{ V}/7.67)\sqrt{2} \cong 19.4 \text{ V}$$

and

$$V_{in(max)} \cong (130 \text{ V}/7.67)\sqrt{2} \cong 23.9 \text{ V}$$

Since $V_{in(min)}$ is larger than the 15-V regulated output V_{out} by more than the 2-V dropout voltage, the regulator should work reliably. In this case, the maximum power dissipation

$$P_{D(max)} \cong (23.9 \text{ V} - 15 \text{ V})200 \text{ mA} = 1.78 \text{ W} \qquad (2\text{-}10)$$

Now we can determine that the sink's thermal resistance should not be larger than

$$\theta_{SA} \cong \frac{150°C - 25°C}{1.78 \text{ W}} - 0.2°C - 3°C = 67°C \qquad (2\text{-}12)$$

REVIEW QUESTIONS

Fill in the blanks of the next ten problems with one of the six following terms:

(I)	infinite	(IV)	relatively small
(II)	zero	(V)	decrease
(III)	relatively large	(VI)	increase

2-1. The ideal Op Amp's output resistance is _____ ohms.

2-2. The ideal Op Amp's open-loop voltage gain is _____ .

2-3. The ideal input resistance of an Op Amp is _____ ohms.

2-4. The ideal Op Amp's bandwidth is _____ hertz.

2-5. The ideal response time of an Op Amp is _____ seconds.

2-6. Compared to the internal resistance of the signal source, the practical Op Amp's effective input resistance should be _____.

2-7. Compared to the load resistance, the effective output resistance of a practical Op Amp should be _____.

2-8. When the differential input voltage is zero, the Op Amp's output voltage ideally should be _____.

2-9. Feedback tends to _____ the effective output resistance and to _____ the effective input resistance.

2-10. Compared to a typical output signal voltage, the differential input should ideally be _____ .

2-11. Comparatively, what do the terms *open-loop voltage gain* and *closed-loop voltage gain* mean referring to Op Amps?

2-12. To which of the Op Amp characteristics does the term *slew rate* apply?

2-13. If the Op Amp's output must be capable of swinging in both positive and negative directions, generally what kind of power supply does it require?

2-14. What does it mean when we say an Op Amp is *nulled*?

2-15. What is the meaning of the term *high CMRR*?

2-16. What undesirable effect might result if an Op Amp's dc source voltages are not well regulated?

2-17. A zener diode has a *(relatively constant)(varying)* voltage across it with a *(relatively constant)(varying)* current through it.

2-18. What is the dropout voltage of a three-terminal regulator?

2-19. What do we call the current a regulator draws from the unregulated supply while the load on the regulator is an open?

PROBLEMS

2-1. Design a half-wave zener-regulated power supply such as the circuit in Fig. 2-6a, with plus and minus 10-V dc outputs. Its loads—a few Op Amps and a relay—will draw currents varying from 10 mA to 50 mA. The transformer is to work off a 117-V line that can vary from 100 V to 130 V rms. Choose a transformer from Table 2-2.

2-2. Design a full-wave zener-regulated power supply, such as the circuit of Fig. 2-6b, with plus and minus 12-V dc outputs. Its loads—a couple of Op Amps, relays and transistor amplifiers—will draw currents varying from 15 mA to 100 mA. The transformer will work off a 117-V line that varies from 105 V to 125 V rms. Choose a transformer from Table 2-2.

2-3. Referring to Example 2-3, can we use the transformer shown in row 2 instead of row 7 of Table 2-2? Explain.

2-4. Referring again to Example 2-3, if a transformer with a 30-V, instead of a 15-V, secondary were used, what would the maximum power dissipation in the regulator be, and what maximum θ_{SA} can the heat sink have?

THE OP AMP WITH
AND WITHOUT FEEDBACK

The extremely high open-loop voltage gain A_{VOL} of the typical Op Amp makes it "touchy" to work with, especially in linear (undistorted) circuit applications. Because of the high gain, a relatively small differential input voltage V_{id} can easily drive an Op Amp's output voltage to its limit. If this occurs, the output signal is clipped and distorted. In this chapter we will see how negative feedback is used with an Op Amp to reduce and stabilize its effective voltage gain. In fact, with feedback we are able to select any specific gain we need as long as it is less than the Op Amp's open-loop gain A_{VOL}.

3.1 OPEN-LOOP CONSIDERATIONS

The Op Amp is generally classified as a linear device. This means that its output voltage V_o tends to proportionally follow changes in the applied differential input V_{id}. Within limits, the changes in output voltage V_o are larger than the changes in the input V_{id} by the open-loop gain A_{VOL} of the Op Amp. The amount that the output voltage V_o can change (swing), however, is limited by the dc supply voltages and the load resistance R_L. Generally, the output voltage swing is restricted to values between the $+V$ and $-V$ supply voltages. As shown in Fig. 3-1, manufacturers provide curves showing their Op Amps' maximum output voltage swing versus supply voltage and also versus load resistance R_L. Note that smaller supply voltages and load resistances reduce an Op Amp's signal output capability.

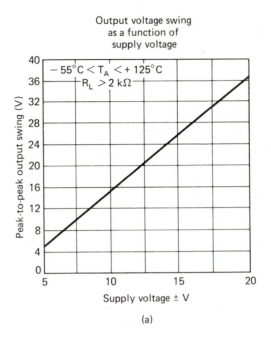

(a)

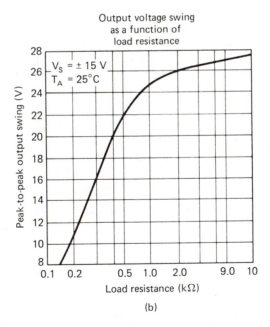

(b)

Figure 3-1   Typical Op Amp characteristics: (a) output voltage V_o swing vs. dc supply voltage; (b) output voltage V_o swing vs. load resistance R_L.

Any attempt to drive the Op Amp beyond its limits results in sharp clipping of its output signals.

For example, the Op Amp in Fig. 3-2 is shown with a small input signal V_{id} applied to its inverting input 1 and the resulting output V_o. Note that the output voltage $V_o = 0$ V at the times when $V_{id} = 0$ V, that is, the circuit is nulled, though for reasons we will see later it is difficult to null the output of an Op Amp if no feedback is used. Note also that the gain $A_{VOL} = 10,000$, which means that the output signal V_o is -10^4 times larger than the input V_{id}, but within limits. In this case, the limits are plus and minus 8 V, as indicated by the fact that the output signal V_o is clipped at $+8$ V on some positive alternations and at -8 V on some negative alternations. Apparently, this Op Amp has output voltage swing vs. supply voltage characteristics as shown in Fig. 3-1. Note that, although the input V_{id} varies somewhat sinusoidally and with relatively low amplitudes, the output V_o is -10^4 times larger and easily reaches the positive and negative limits of the Op Amp. Consequently severe clipping occurs. Only the smaller input signals—1.6 mV peak to peak or less—are faithfully reproduced. There are applications where the Op Amp is purposely used as a signal-squaring circuit in which deliberately driving the Op Amp's input with a relatively large signal results in almost a squarewave output. This output appears more like a square and less like a trapezoid when the input signals have larger amplitudes and lower frequencies.

If the same input signal is applied to the noninverting input 2, as shown in Fig. 3-3, the output is $+10^4$ times larger, within limits of course. In this case, the output swings positively and then negatively on the positive and negative alternations of the input signal respectively. Clipping occurs when the output attempts to exceed the Op Amp's limits, just as in the inverting mode.

Example 3-1

Sketch output voltage waveforms of an Op Amp, such as in Fig. 3-2, if the input signal is as shown but if:

(a) the open-loop gain $A_{VOL} = 5000$.
(b) the open-loop gain $A_{VOL} = 100,000$

Answer. See Fig. 3-4. Note that with a larger gain, A_{VOL}, a given input signal tends to be more distorted. This does not mean that a high open-loop gain is undesirable, on the contrary, the larger A_{VOL} is the better. As we will see, clipping can easily be controlled with feedback.

3.2 FEEDBACK AND THE INVERTING AMPLIFIER

Some form of feedback is usually used with Op Amps in linear applications. Negative feedback enables the Op Amp circuit designer to select and control voltage gain easily. Generally, an amplifier has negative feedback if a

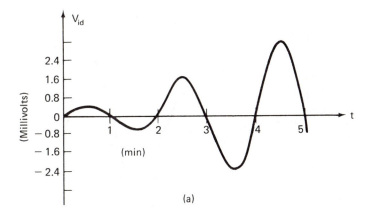

(a)

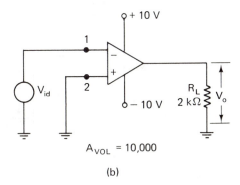

$A_{VOL} = 10,000$

(b)

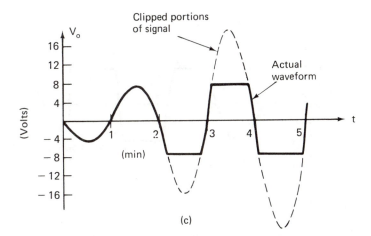

(c)

Figure 3-2   (a) Differential input signal; (b) Op Amp with no feedback (inverting mode); (c) output signal waveform if $A_{VOL} = 10^4$.

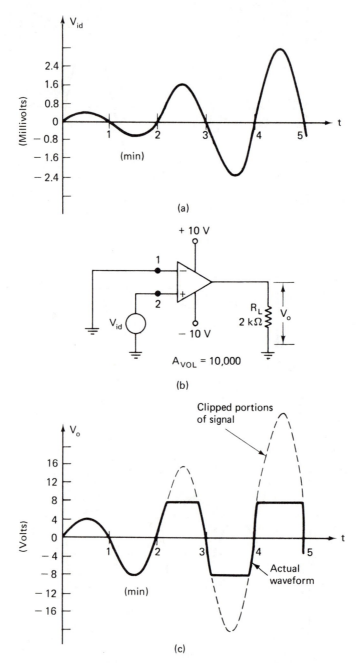

Figure 3-3   (a) Differential input signal; (b) noninverting Op Amp circuit with no feedback; (c) output signal waveform.

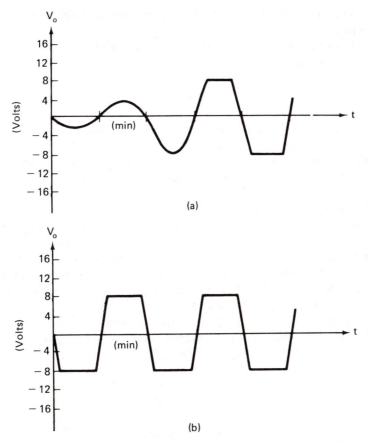

Figure 3-4 (a) Output voltage of the Op Amp in Fig. 3-2 if $A_{VOL} = 5000$; (b) output voltage of the Op Amp in Fig. 3-2 if $A_{VOL} = 100,000$.

portion of its output is fed back to its input and if the input and output signals are out of phase. The Op Amp circuit in Fig. 3-5 has negative feedback. Note that a resistor R_F is across the Op Amp's output and inverting input terminals. This, along with R_1, causes a portion of the output signal V_o to be fed back to the inverting input terminal. With this type of feedback, the

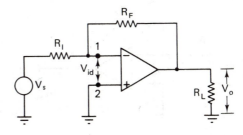

Figure 3-5 Op Amp connected to work as an inverting amplifier with feedback.

circuit's effective voltage gain A_v is typically much smaller than the Op Amp's open-loop gain A_{VOL}.

In the circuit shown in Fig. 3-5, the input signal is voltage V_s; therefore, its effective voltage gain is

$$A_v = \frac{V_o}{V_s} \qquad (3\text{-}1)$$

This ratio of output signal voltage V_o to the input signal V_s is also called the *closed-loop gain* because it is the gain when a feedback resistor R_F completes a loop from the amplifier's output to its inverting input 1. The specific value of A_v is determined mainly by the values of resistors R_1 and R_F.

We can see how resistors R_1 and R_F in the circuit of Fig. 3-5 determine the closed-loop gain A_v if we analyze this circuit's current paths and voltage drops. For example, as shown in Fig. 3-6, the signal source V_s drives a current I through R_1. Assuming that the Op Amp is ideal (having infinite resistance looking into input 1), all of this current I flows up through R_F. By Ohm's law we can show that the voltage drop across R_1 is $R_1 I$ and that the voltage across R_F is $R_F I$. The significance of this will be seen shortly.

Since the open-loop gain A_{VOL} of the Op Amp is ideally infinite or at least very large practically, the differential input voltage V_{id} is infinitesimal compared to the output voltage V_o. Thus since

$$A_{VOL} = \frac{V_o}{V_{id}} \qquad (2\text{-}3a)$$

then

$$V_{id} = \frac{V_o}{A_{VOL}}$$

This last equation shows that, the larger A_{VOL} is with any given V_o, the smaller V_{id} must be compared to V_o. The point is, the voltage V_{id} across inputs 1 and 2 is practically zero because of the large open-loop gain A_{VOL} of the typical Op Amp. Therefore, with *virtually* no potential difference between inputs 1 and 2, and with the noninverting input 2 grounded, the inverting input is *virtually* grounded too. This circuit's equivalent can therefore be shown as in Fig. 3-7. With voltage V_o to ground at the right of

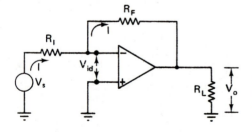

Figure 3-6 Currents in R_1 and R_F are equal if the Op Amp is ideal; they are approximately equal with practical Op Amp if R_F is not too large.

$$V_o \cong R_F I$$

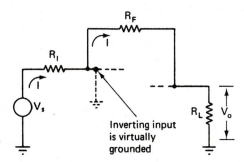

Figure 3-7 Equivalent circuit of Figs.
3-5 and 3-6; $V_s \cong R_1 I$ and $V_o \cong R_F$.

Inverting input
is virtually
grounded

R_F and virtual ground at its left, the voltage across R_F is V_o for practical purposes. By Ohm's law.

$$V_o \cong R_F I \qquad (3\text{-}2)$$

Similarly, we can see that the voltage V_s to ground is applied to the left of R_1, while its right end is virtually grounded. Therefore, for most practical purposes, the voltage across R_1 is V_s. Again with Ohm's law we can show that

$$V_s \cong R_1 I \qquad (3\text{-}3)$$

Since the closed-loop gain A_v is the ratio of the output signal voltage V_o to the input signal voltage V_s, we can substitute Eqs. (3-2) and (3-3) into this ratio and show that

$$A_v = \frac{V_o}{V_s} \cong -\frac{R_F I}{R_1 I} \cong -\frac{R_F}{R_1} \qquad (3\text{-}4)$$

The negative sign means that the input and output signals are out of phase. This last equation shows that by selecting a ratio of feedback resistance R_F to the input resistance R_1, we select the inverting amplifier's closed-loop gain A_v.

There are limits on the usable values of R_F and R_1 which are caused by practical design problems. Though these design problems are discussed in later chapters, for the present we should know that the feedback resistance R_F is rarely larger than 10 MΩ. More frequently, R_F is 1 MΩ or less. Since one end of the signal source V_s is grounded and one end of R_1 is virtually grounded, V_s sees R_1 as the amplifier's input resistance. To avoid loading of (excessive current drain from) the signal source V_s, R_1 is typically 1 kΩ or more.

Example 3-2

Referring to the circuit in Fig. 3-8a, find its voltage gain V_o/V_s when the switch S is in:
(a) position 1,
(b) position 2, and

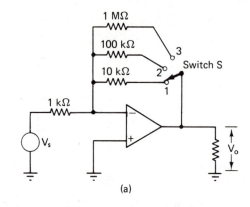

(a)

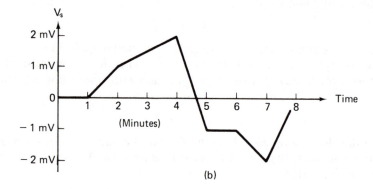

(b)

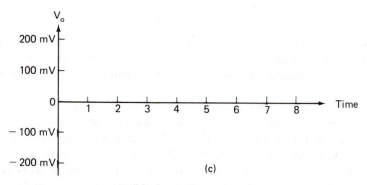

(c)

Figure 3-8

(c) position 3.

(d) If the switch is in position 2 and the input voltage V_s has the waveform shown in Fig. 3-8b, sketch the output voltage waveform V_o. Assume that when V_s was zero, the output V_o was zero (nulled).

(e) What is the resistance seen by the signal source V_s for each of the switch positions?

Answer.   (a) When the switch S is in position 1, the feedback resistance $R_F = 10\,\text{k}\Omega$. The input resistor $R_1 = 1\,\text{k}\Omega$ regardless of the switch position. Therefore, the gain is

$$A_v \cong -\frac{R_F}{R_1} = -\frac{10\,\text{k}\Omega}{1\,\text{k}\Omega} = -10 \tag{3-4}$$

(b) With the switch S in position 2, $R_F = 100\,\text{k}\Omega$; therefore

$$A_v \cong -\frac{100\,\text{k}\Omega}{1\,\text{k}\Omega} = -100$$

(c) With the switch S in position 3, $R_F = 1\,\text{M}\Omega$; therefore,

$$A_v \cong -\frac{1\,\text{M}\Omega}{1\,\text{k}\Omega} = -1000$$

The negative signs mean that the input and output signal voltages are out of phase.

(d) Since the gain $A_v = -100$ with the switch S in position 2, the input signal voltage V_o is amplified by 100 and its phase is inverted, resulting in an output voltage waveform shown in Fig. 3-9. By rearranging Eq. (3-1), we can show that

$$V_o = A_v V_s$$

Thus, at $t = 2$ min, $V_o = -100\,(1\,\text{mV}) = -100\,\text{mV}$. At $t = 4$ min, $V_o = -100\,(2\,\text{mV}) = -200\,\text{mV}$, etc.

(e) The signal source V_s sees R_1 as the load regardless of the switch position.

Example 3-3

The Op Amp in Fig. 3-8 operates with dc supply voltages of ± 15 V and with a load resistance $R_L = 200\,\Omega$. If the characteristics shown in Fig. 3-1 are typical of this Op Amp, what are the values of:

(a) the maximum possible peak-to-peak unclipped output signal V_o, and

(b) the maximum input peak-to-peak signal voltage V_s that can be applied and not cause clipping of V_o while the switch S is in position 2?

Answer.   (a) Referring to Fig. 3-1b, we project up from $0.2\,\text{k}\Omega$ to the curve. Directly to the left of the intersection of our projection and curve, we see that this amplifier's

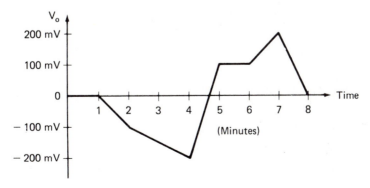

Figure 3-9

peak-to-peak output swing capability is about 11 V. This means that clipping will occur if we attempt to drive the output V_o beyond $+5.5$ V or -5.5 V.

(b) With the switch S in position 2, the gain is -100 as we found in the previous problem. Since the maximum peak-to-peak output swing is about 11 V, the maximum peak-to-peak input swing is determined as follows. Since

$$A_v = \frac{V_o}{V_s} \tag{3-1}$$

then

$$V_s = \frac{V_o}{A_v} \cong \frac{11 \text{ V(p-p)}}{-100} = |110 \text{ mV(p-p)}|$$

3.3 FEEDBACK AND THE NONINVERTING AMPLIFIER

We can use the Op Amp, with feedback, in a noninverting mode much as we used it in the inverting mode discussed in the previous section. We can drive the noninverting input 2 with a signal source V_s instead of indirectly driving the inverting input 1 as shown in Figs. 3-3 and 3-10. As with the inverting amplifier, the values of externally connected resistors R_1 and R_F determine the circuit's closed-loop voltage gain A_v. A gain equation in terms of R_1 and R_F can be worked out if we analyze the noninverting amplifier's currents and voltages.

As shown in Fig. 3-11a, the signal current I is the same through resistors R_1 and R_F as long as the resistance R_i looking into the inverting input 1 is infinite or at least very large. In other words, R_1 and R_F are effectively in series. As before, the voltage drops across these resistors can be shown as R_1I and R_FI. At the right side of R_F we have the output voltage V_o to ground. This voltage is also across R_1 and R_F because these resistors are effectively in series and the left side of R_1 is grounded. Thus, we can show that the output voltage V_o is the sum of the drops across R_1 and R_F, that is

$$V_o \cong R_1I + R_FI$$

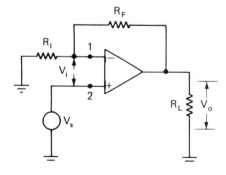

Figure 3-10 Op Amp wired to work as a noninverting amplifier (noninverting mode).

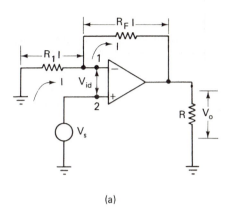

(a)

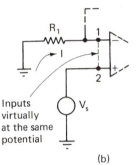

Figure 3-11 (a) Noninverting amplifier with currents and voltage drops shown; (b) the difference voltage V_{id} across inputs 1 and 2 is so small compared to V_s that these inputs are virtually shorted.

Inputs virtually at the same potential

(b)

Factoring I out, we get

$$V_o \cong (R_1 + R_F)I \qquad (3\text{-}5)$$

As with the inverting amplifier, the differential input voltage V_{id} is zero for most practical purposes. Due to the very large open-loop gain A_{VOL} of the typical Op Amp, we can assume that there is practically no potential difference between points 1 and 2 as shown in Fig. 3-11b. Therefore, nearly all of the input signal voltage V_s appears across R_1, and by Ohm's law, we can still show that

$$V_s \cong R_1 I \qquad (3\text{-}3)$$

The voltage gain A_v of the noninverting amplifier is the ratio of its output V_o to its input V_s. Thus, substituting the right sides of Eqs. (3-3) and (3-5) into this ratio yields

$$A_v = \frac{V_o}{V_s} \cong \frac{(R_F + R_1)I}{R_1 I} = \frac{R_F + R_1}{R_1} = \frac{R_F}{R_1} + 1 \qquad (3\text{-}6)$$

Since the resistance seen looking into the noninverting input is infinite if the Op Amp is ideal (or at least very large with a practical Op Amp), the signal source V_s sees a very large resistance looking into the Op Amp of Fig. 3-10. Generally, the effective input resistance of the noninverting amplifier is larger with circuits wired to have lower closed-loop gains. More specifically, the effective opposition to signal current flow into the noninverting input is an impedance $Z_{i(eff)}$ instead of a resistance when input capacitances are considered. Looking into the noninverting input 2, the signal current sees parallel paths: through an impedance Z_{ic} and through the Op Amp's resistance R_i (see Fig. 3-12). Thus the total effective input impedance is

$$Z_{i(eff)} \cong \frac{1}{A_v/A_{VOL}R_i + 1/Z_{ic}} \qquad (3\text{-}7a)$$

where R_i is the input resistance measured between input 1 and 2, open-loop,
   Z_{ic} is the impedance to ground or common measured from either input,
   A_{VOL} is the open-loop gain, and
   A_v is the closed-loop gain.

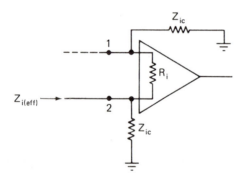

Figure 3-12 Equivalent circuit of an Op Amp; R_i is the resistance between inputs 1 and 2 specified by the manufacturer; Z_{ic} is the impedance to ground or common from either input which is actually internal to the Op Amp.

Typically, Z_{ic} is much larger than R_i, especially at low frequencies. If we assume that Z_{ic} is relatively very large, Eq. (3-7a) can be simplified to

$$Z_{i(eff)} \cong \frac{1}{A_v/A_{VOL}R_i} = \left(\frac{A_{VOL}}{A_v}\right)R_i \qquad (3\text{-}7b)$$

where the ratio A_{VOL}/A_v is called the *loop gain*. Note that if the Op Amp is wired to have a low closed-loop gain A_v, say approaching 1, the effective input impedance $Z_{i(eff)}$ theoretically approaches a value that is A_{VOL} times larger than the manufacturer's specified R_i.

Example 3-4

If $R_1 = 1 \text{ k}\Omega$ and $R_F = 10 \text{ k}\Omega$ in the circuit of Fig. 3-10, what is the voltage gain V_o/V_s? If this circuit's input signal voltage V_s has the waveform shown in Fig. 3-13a, sketch the resulting output V_o. Assume that the Op Amp was initially nulled.

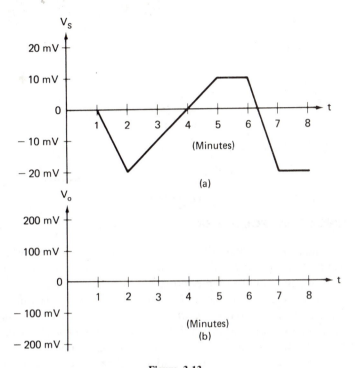

Figure 3-13

Answer.   Since this Op Amp is wired in a noninverting mode, the voltage gain can be determined with Eq. (3-6):

$$A_v \cong \frac{R_F}{R_1} + 1 = 11$$

Therefore, the output voltage is 11 times larger than, and in phase with, the input signal V_s. Thus at $t = 2$ minutes, $V_o = 11(-20 \text{ mV}) = -220 \text{ mV}$. Similarly, at $t = 5$ minutes, $V_o = 11(10 \text{ mV}) = 110 \text{ mV}$, etc. (see Fig. 3-14 for the complete output waveform.)

Example 3-5

Refer to the circuit described in Example 3-4. If the Op Amp is a 741C type as listed in Appendix F, what approximate minimum impedance does the signal source V_s see if the impedance to ground from either input is assumed to be infinite?

Answer.   As shown in Appendix F, the 741C's minimum A_{VOL} and R_i values are 20,000 and 150 kΩ, respectively. Since the circuit is wired externally for a closed-loop gain of 11, we can find the minimum effective impedance seen by V_s with Eq. (3-7b). Thus in this case

$$Z_{i(\text{eff})} \cong \left(\frac{20{,}000}{11}\right) 150 \text{ k}\Omega \cong 273 \text{ M}\Omega \qquad (3\text{-}7b)$$

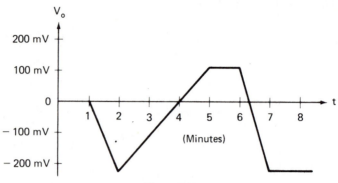

Figure 3-14

3.4 THE VOLTAGE FOLLOWER

In some applications, an amplifier's voltage gain is not as important as its ability to match a high internal resistance signal source to a low, possibly varying, resistance load. The Op Amp in Fig. 3-15 is connected to work as a *voltage follower* which has an extremely large input resistance and is capable of driving a relatively low-resistance load. The voltage follower's output resistance is very small; therefore, variations in its load resistance negligibly affect the amplitude of the output signal. When used between a high-internal-resistance signal source and a smaller, varying-resistance load, the voltage follower is called a *buffer amplifier*.

The voltage follower is simply a noninverting amplifier, similar to the one in Fig. 3-11a, where R_1 is replaced with infinite ohms (an open) and R_F is replaced with zero ohms (a short). (See Fig. 3-16.) Due to the Op Amp's large open-loop gain A_{VOL}, the differential input voltage V_{id} is very small; therefore, the inputs 1 and 2 are virtually at the same potential. Since the output signal V_o is the voltage at input 1, and since the input signal V_s is directly applied to input 2, then

$$V_o \cong V_s$$

Apparently, with the input and output signal voltages nearly equal, the voltage gain of the voltage follower is very nearly 1 (unity). We can see this

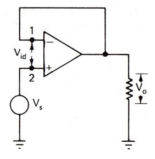

Figure 3-15 Op Amp connected to work as a voltage follower.

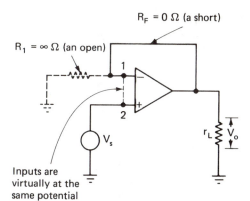

Figure 3-16 The voltage follower of Fig. 3-10 can be shown to be a noninverting amplifier.

another way—by substituting $R_F = 0\ \Omega$ into the gain Eq. (3-6). The term *voltage follower* therefore describes the circuit's function: the output voltage V_o follows the input voltage V_s waveform.

As with the noninverting amplifier, the effective input impedance $Z_{in(eff)}$ of the voltage follower is about the loop gain times larger than R_i, where R_i is the differential input resistance measured under open-loop conditions. Since the voltage follower's closed-loop gain $A_v \cong 1$, its loop gain A_{VOL}/A_v is very large—about equal to A_{VOL}. This accounts for the extremely high input impedance of the voltage follower.

As the input impedance $Z_{in(eff)}$ is made more ideal (increased) by the use of negative feedback, the output resistance is also improved (reduced). More specifically, the effective output resistance $R_{o(eff)}$ of an Op Amp is smaller than the output resistance R_o measured with open-loop conditions by the loop-gain factor. That is,

$$R_{o(eff)} \cong \left(\frac{A_v}{A_{VOL}}\right) R_o \qquad (3\text{-}8)$$

This shows that, the smaller the closed-loop gain A_v, the smaller the effective output resistance $R_{o(eff)}$.

Example 3-6

(a) If the circuit in Fig. 3-15 has the input voltage waveform V_s shown in Fig. 3-9, sketch the output voltage waveform. Assume that initially the Op Amp was nulled.

(b) If a signal source has 100 kΩ of internal resistance and an output of 4 m V *before* being connected to input 2 (open circuited), what is the output of the signal source *after* it is connected to the input 2 of this circuit?

Answer. (a) Since the gain A_v of the voltage follower is unity, and since there is no phase inversion, the output V_o has the same waveform as the input V_s.

(b) The input impedance of the voltage follower is in the range of hundreds of megohms and therefore appears as an open compared to the 100 kΩ internal resistance of the signal source. Thus, there is essentially no signal voltage drop across the internal resistance, and the output of the signal generator remains very close to its 4-mV open-circuit value after being connected to input 2. Therefore the output of this voltage follower is also about 4 m V.

REVIEW QUESTIONS

3-1. What two factors outside of the Op Amp affect its maximum unclipped output signal capability?

3-2. When the Op Amp is connected as an inverting amplifier (Fig. 3-5), what is the approximate input resistance as seen by the signal source V_s?

3-3. Since the Op Amps's open-loop gain is usually very high, why is it that the Op Amp is seldom used as a signal amplifier in open loop?

3-4. If a low-frequency sine wave with a 1-V peak-to-peak amplitude were directly applied to an Op Amp's inputs 1 and 2, what kind of waveform would you expect at the output?

3-5. What is the difference between the closed-loop gain and the open-loop gain?

3-6. How does the differential input signal V_{id} compare in magnitude to the output voltage V_o?

3-7. How can an Op Amp input terminal be virtually grounded but not actually grounded?

3-8. How does the input resistance of a noninverting amplifier compare with the input resistance of the inverting type?

3-9. If an Op Amp is described as having been nulled, what does this mean?

3-10. What effect does negative feedback have on the voltage gain of an amplifier?

3-11. If an Op Amp is connected to have negative feedback, how does its effective output resistance compare with the output resistance specified by the manufacturer?

3-12. If an Op Amp is connected to operate as a noninverting amplifier, with a closed-loop gain lower than the open-loop gain, how does its effective input impedance compare with the input resistance specified by the manufacturer?

3-13. Which would you expect to have larger input impedance: an Op Amp connected to work as a noninverting amplifier with a closed-loop gain of 101, or the same Op Amp connected to work as a voltage follower?

3-14. Since the voltage follower has only unity voltage gain, what good is it?

3-15. What might happen to the output signal voltage waveform if the load resistance is too small?

3-16. What might happen to the output signal voltage waveform if the dc supply voltages are too small?

PROBLEMS

In the circuit of Fig. 3-17a, the signal driving the Op Amp is taken off a bridge circuit as shown. The thermistor in the bridge is a resistance with a high negative temperature coefficient, that is, its resistance significantly decreases or increases as

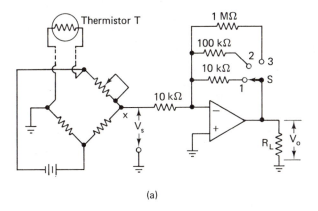

(a)

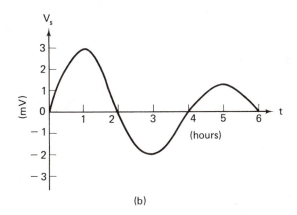

(b)

Figure 3-17

its temperature increases or decreases, respectively. The thermistor is in an oven. As the oven's temperature rises or falls, the voltage V_s swings positively or negatively. The load R_L represents the effective resistance of a motor's armature. When the shaft of the motor turns, it operates a fuel valve and thus admits more or less fuel to the oven. This system therefore controls the temperature in the oven. The switch S provides a means of selecting its sensitivity (closed loop voltage gain).

If the voltage across the bridge, which is voltage V_s, has the waveform shown in Fig. 3-17b, find the voltage gain A_v and select a waveform from Fig. 3-18 that best represents the Op Amp's output V_o for each of the following conditions:

3-1. Switch S is in position 1.

3-2. Switch S is in position 2.

3-3. Switch S is in position 3.

3-4. What is the load resistance across the bridge (points x and ground) in the circuit of Fig. 3-17a?

If the Op Amp circuit is removed from the system and replaced with the one in

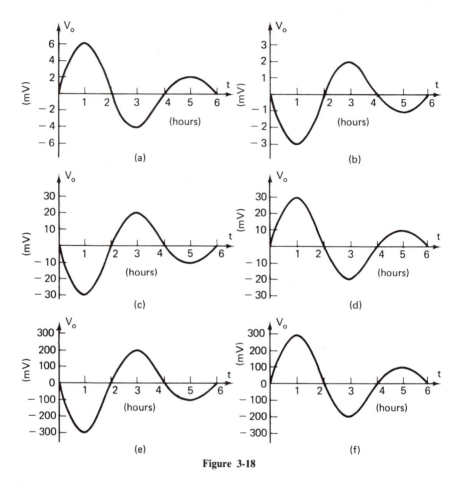

Figure 3-18

Fig. 3-19, but the waveform V_s is still as shown in Fig. 3-17b, find this circuit's voltage gain and select a waveform from Fig. 3-18 that best represents the Op Amp's output V_o for each of the following conditions:

3-5. The switch S is in position I.

3-6. The switch S is in position II.

3-7. The switch S is in position III.

3-8. Which of the Op Amp circuits—the one in Fig. 3-17a or the one in Fig. 3-19— draws the larger current from the bridge?

3-9. Which of the Op Amp circuits—the one in Fig. 3-17a or the one in Fig. 3-19— offers the more constant load on the bridge (keep in mind that the switch S position might frequently be changed)?

    The switches S_1 and S_2 in the circuit of Fig. 3-20 are ganged, that is, when S_1 is in position 1, S_2 is also in position 1. Similarly, S_1 and S_2 are in position 2 simultaneously.

3-10. If the signal source is 2 V dc, R_s = 50 kΩ, and R_L = 1 kΩ in the circuit of Fig. 3-

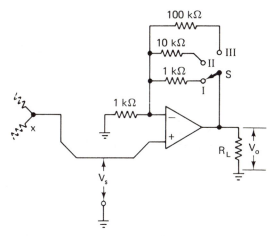

Figure 3-19

20, what is the approximate voltage at point 0 to ground when S_1 is in position 2?

3-11. If the signal source is 3 V dc, $R_s = 150$ kΩ, and $R_L = 1$ kΩ in the circuit of Fig. 3-20, what is the approximate voltage at point 0 with respect to ground when S_1 is in position 2?

3-12. Referring to Prob. 3-10, what is the voltage at point 0 with respect to ground when the switches are in position 1?

3-13. Referring to Prob. 3-11, what is the voltage at point 0 with respect to ground when the switches are in position 1?

3-14. What is the typical effective output resistance of the 709 Op Amp whose characteristics are given in Appendix H, when it is connected to work as a voltage follower?

3-15. What is the typical effective output resistance of the 741 Op Amp whose characteristics are given in Appendix I, when it is connected to work with a closed-loop gain A_v of 1000?

3-16. Over what range can the gain A_v of the circuit shown in Fig. 3-21a be adjusted if the resistor R can be varied from 0 Ω to 1 MΩ?

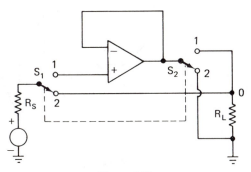

Figure 3-20

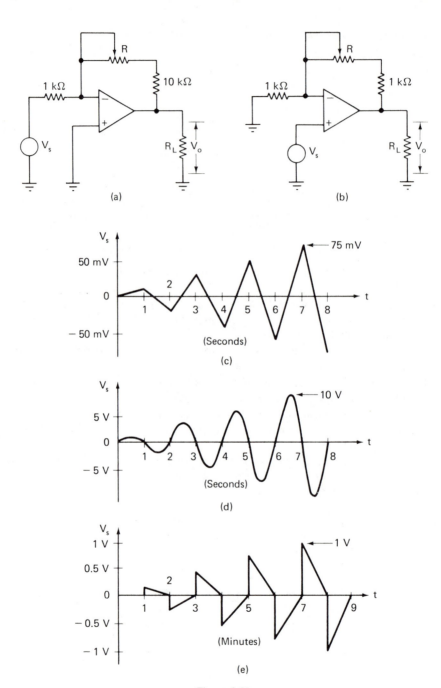

Figure 3-21

3-17. If $R = 190$ kΩ in the circuit of Fig. 3-21a, what is the output voltage V_o if $V_s = -2$ m V dc? Assume V_o was OV when V_s was OV.

3-18. If $V_s = -3$ m V dc and $R = 5$ kΩ in the circuit shown in Fig. 3-21b, what is the output voltage V_o?

3-19. If resistance R can be adjusted from 0 Ω to 200 kΩ in the circuit of Fig. 3-21b, over what range can its voltage gain A_v be varied?

The following four problems refer to the circuit in Fig. 3-21a with the input voltage waveform as shown in Fig. 3-21c. The Op Amp was initially nulled and has the characteristics in Fig. 3-1.

3-20. Sketch the output waveform V_o if the dc supply voltages are $+15$ V and -15 V, the resistance R is adjusted to 90 kΩ, and $R_L = 2$ kΩ.

3-21. Sketch the output waveform V_o if the dc supply voltages are ±15 V, the resistance R is adjusted to 140 kΩ, and $R_L = 1$ kΩ

3-22. Referring to Prob. 3-20, sketch the output waveform V_o if the dc supply voltages are reduced from ±15 V to ±7.5V. The R and R_L values are unchanged.

3-23. Referring to Prob. 3-21, sketch the output waveform V_o if the load resistance R_L is reduced to 200 Ω

The following three problems refer to the circuit in Fig. 3-21b with the input voltage as shown in Fig. 3-21d. The Op Amp was initially nulled and has the characteristics in Fig. 3-1.

3-24. Sketch the output waveform V_o if the dc supply voltages are ±15 V, the resistance R is adjusted to 0 Ω and $R_L = 2$ kΩ.

3-25. Sketch the output waveform V_o if the dc supply voltages are ±5 V, the resistance R is adjusted to 0 Ω, and $R_L = 2$ kΩ.

3-26. Sketch the output waveform V_o if the dc supply voltages are ±15 V, the resistance R is adjusted to 0 Ω and $R_L = 200$ Ω.

3-27. Referring to Fig. 3-21a, if the circuit has the input waveform shown in Fig. 3-21e, $R = 40$ kΩ, $R_L = 2$ kΩ, ±15-V supply voltages, and the Op Amp characteristics in Fig. 3-1, show the output voltage waveform.

3-28. Rework the previous problem using the circuit in Fig. 3-21b instead. The values of R, R_L, and supply voltages are unchanged.

4

OFFSET
CONSIDERATIONS

The practical Op Amp, unlike the hypothetical ideal version, has some dc output voltage, called *output offset* voltage, even though both of its inputs are grounded. Such an output offset is an error voltage and is generally undesirable. The causes and cures of output offset voltages are the subjects of this chapter. Here we will become familiar with parameters that enable us to predict the maximum output offset voltage that a given Op Amp circuit can have. On the foundation laid in this chapter, we will build an understanding of why, and an ability to predict how much, a given Op Amp's output voltage tends to drift with power supply and temperature changes which are important subjects discussed later.

4.1 INPUT OFFSET VOLTAGE V_{io}

The input offset voltage V_{io} is defined as the amount of voltage required across an Op Amp's inputs 1 and 2 to force the output voltage to 0 V. In a previous chapter we learned that ideally, when the differential input voltage $V_{id} = 0$ V, as in Fig. 4-1, the output offset V_{oo} is 0 V too. In the practical case, however, some V_{oo} voltage will usually be present. This is caused by imbalances within the Op Amp's circuitry. They occur because the transistors of the input differential stage within the Op Amp usually admit slightly different collector currents even though both bases are at the same potential. This causes a differential output voltage from the first stage, which is amplified and possibly aggravated by more imbalances in the following

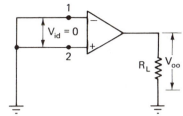

Figure 4-1   Ideally, $V_{oo} = 0$.

stages. The combined result of these imbalances is the output offset voltage V_{oo}. If small dc voltage of proper polarity is applied to inputs 1 and 2, it will decrease the output voltage. The amount of differential input voltage, of the correct polarity, required to reduce the output to 0 V is the input offset voltage V_{io}. When the output is forced to 0 V by the proper amount and polarity of input voltage, the circuit is said to be *nulled* or *balanced*. The polarity of the required input offset voltage V_{io} at input 1 with respect to input 2 might be positive as often as negative. This means that the output offset voltage, before nulling, can be positive or negative with respect to ground.

Common nulling circuits of inverting-mode amplifiers are shown in Fig. 4-2. Their equivalents for noninverting amplifiers are shown in Fig. 4-3. In each of these circuits, the potentiometer POT can be adjusted to provide the value and polarity of dc input voltage required to null the output to 0 V. As shown, the ends of the POT are connected to the positive and negative dc supply voltages. With some Op Amps, such as types 748, 777, 201, 741,* etc. (see Appendix E), *offset adjust* pins are provided. The manufacturers of 741s recommend that a POT be placed across their OFFSET NULL pins as

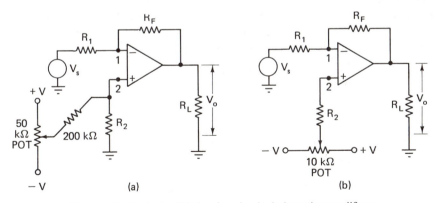

Figure 4-2   Typical null balancing circuits in inverting amplifiers.

* Generally, 741-type Op Amps have characteristics such as listed in Appendix I. 741s are identified in several ways, depending on the manufacturer. To name a few: Fairchild μA741, Motorola MC1741, National Semiconductor LM741, Raytheon RM741, Signetics, μA741.

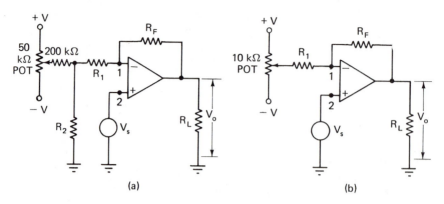

Figure 4-3 Typical null balancing circuits in noninverting amplifiers.

shown in Fig. 4-4. Adjustment of this POT will null the output if the closed loop gain A_v is not too large.

If an Op Amp circuit is not nulled, more or less output offset voltage V_{oo} exists, depending on its specified input offset voltage V_{io}, closed-loop gain A_v, and other factors discussed in the following sections. When the source voltage $V_s = 0$ V, an equivalent circuit of *either* the inverting *or* the noninverting amplifier is as shown in Fig. 4-5. As shown in this figure, the Op Amp's specified input offset voltage V_{io} is equivalent to a dc signal source working into a noninverting type amplifier. If this equivalent circuit in Fig. 4-5 represents an inverting amplifier, the signal source V_s is replaced with its own internal resistance at point x. If this equivalent circuit represents a noninverting amplifier, V_s is replaced with its own internal resistance at point y. A short circuit replaces V_s if its internal resistance is negligible. According to Fig. 4-5 then, we can show that the output offset voltage V_{oo}, "caused by" the input offset voltage, is the product of the closed-loop gain and the specified V_{io}:

$$V_{oo} = A_v V_{io} \tag{4-1}$$

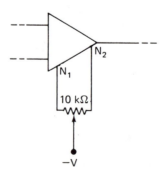

Figure 4-4 Null adjustment on an Op Amp type that has null adjustment pins provided.

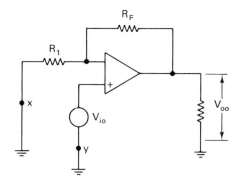

Figure 4-5 Equivalent circuit of either the inverting or the noninverting amplifier.

where

$$A_v \cong \frac{R_F}{R_1} + 1 \qquad (3\text{-}6)$$

Example 4-1

If the Op Amp in the circuit of Fig. 4-6 is a 741 type (see Appendix I), what is the *maximum* possible output offset voltage V_{oo}, caused by the input offset voltage V_{io}, before any attempt is made to null the circuit with the 10-kΩ POT?

Answer.   According to the specifications of the 741, its maximum $V_{io} = 6$ mV. Since the input resistance $R_1 = 1$ kΩ, assuming that the internal resistance of V_s is negligible and since the feedback resistance $R_F = 100$ kΩ, the input offset voltage is multiplied by

$$A_v \cong \frac{R_F}{R_1} + 1 = \frac{100 \text{ k}\Omega}{1 \text{ k}\Omega} + 1 = 101$$

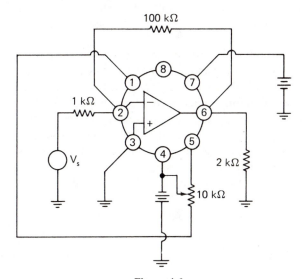

Figure 4-6

Therefore, before the circuit is nulled, we might have an output offset as large as

$$V_{oo} = A_v V_{io} \cong 101(6\text{mV}) = 606 \text{ mV} \quad \text{or about 0.6 V}$$

This means that the output voltage with respect to ground can be either a positive or a negative 0.6 V even though the input signal $V_s = 0$ V.

4.2 INPUT BIAS CURRENT I_B

Most types of IC Op Amps have two transistors in the first (input) differential stage. Transistors, being current-operated devices, require some base bias currents. Therefore, small dc bias currents flow in the input leads of the typical Op Amp as shown in Fig. 4-7. An input bias current I_B is usually specified on the Op Amp specification sheets as shown in the Appendices. It is defined as the average of the two base bias currents, that is,

$$I_B = \frac{I_{B_1} + I_{B_2}}{2} \tag{4-2}$$

These base bias currents, I_{B_1} and I_{B_2}, are *about* equal to each other, and therefore the specified input bias current I_B is *about* equal to either one of them, that is,

$$I_B \cong I_{B_1} \cong I_{B_2} \tag{4-3}$$

Depending on the type of Op Amp, the value of input bias current is usually small, generally in the range from a few to a few hundred nanoamperes in general-purpose, economical Op Amps. Though this seems insignificant, it can be a problem in circuits using relatively large feedback resistors, as we will see.

A difficulty that the base bias currents might cause can be seen if we analyze their effect on a typical amplifier such as in Fig. 4-8. Note that I_{B_2} flows out of ground directly into input 2. (See Fig. 1-8 for the current paths of a typical differential input stage.) Therefore, input 2 is 0 V with respect to ground. On the other hand, base bias current I_{B_1} sees resistance on its way to input 1, that is, the parallel paths containing R_1 and R_F have a total effective resistance as seen by the base bias current I_{B_1}. The flow of I_{B_1} through this effective resistance causes a dc voltage to appear at input 1. With a dc voltage to ground at input 1 while input 2 is 0 V to ground, a dc differential

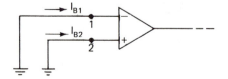

Figure 4-7 Base bias currents flow into the Op Amp and return to ground through the power supplies.

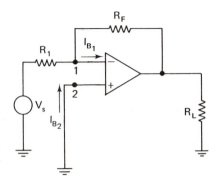

Figure 4-8 Current I_{B_1} sees some resistance on its way to ground, whereas I_{B_2} sees a short.

input voltage V_{id} appears *across* these inputs. This dc input amplified by the closed-loop gain causes an output offset voltage V_{oo}. The amount of output offset voltage V_{oo}, caused by the base bias current I_{B_1}, can be approximated with the equation*

$$V_{oo} \cong R_F I_B \qquad (4\text{-}4)$$

where $I_B \cong I_{B_1}$, and is usually specified by the manufacturer.

Example 4-2

Referring to the circuit in Fig. 4-6, what maximum output offset voltage might it have, caused by the base bias current, before it is nulled? The Op Amp is a type 741.

Answer. According to the 741's specification sheets (Appendix I), the maximum input bias current $I_B = 500$ nA. Since $R_F = 100$ kΩ, the base bias current could cause an output offset as large as

$$V_{oo} \cong R_F I_B = 100 \text{ k}\Omega(500 \text{ nA}) = 50 \text{ mV} \qquad (4\text{-}4)$$

We should note in these last two examples that the output offset caused by the input offset voltage V_{io} is about ten times larger than the output offset caused by the input bias current I_B. In this case then, the input offset voltage V_{io} is potentially the greater problem. However, according to Eq. (4-4), we can see that the output offset V_{oo} caused by the bias current I_B is larger, and therefore potentially more troublesome, with larger values of feedback resistance R_F.

Example 4-3

Referring again to the circuit of Fig. 4-6, suppose that we replace the 1 kΩ input resistor with 100 kΩ and the 100 kΩ feedback resistor with 10 MΩ:

   (a) How does this affect the closed-loop gain A_v of this circuit compared to what it was originally? Before any attempt is made to null this circuit, find
   (b) Its maximum output offset caused by the input offset voltage V_{io}, and
   (c) The maximum output offset caused by the input bias current I_B? The Op Amp is still a type 741.

Answer.   (a) The closed-loop gain A_v is unchanged, that is, this circuit's gain is very nearly equal to the ratio R_F/R_1 which was not changed.

   * See Appendix J for the derivation.

(b) With no change in the closed-loop gain, the output offset caused by the input offset voltage does not change (see Eq. (4-1)).

(c) The output offset caused by input bias current I_B is larger with circuits using larger feedback resistors, according to Eq. (4-4). Thus in this case

$$V_{oo} \cong R_F I_B = 10 \text{ M}\Omega(500 \text{ nA}) = 5 \text{ V}$$

This is a relatively large output offset. It approaches the maximum swing capability of some Op Amp circuits, especially if fairly small dc supply voltages or load resistance R_L are used. This example presents a good case of the use of small feedback resistors.

The effect of the input bias current I_B on the output offset voltage can be minimized if a resistor R_2 is added in series with the noninverting input as shown in Fig. 4-9. By selecting the proper value of R_2, we can make the resistance seen by the base bias current I_{B_2} equal to the resistance seen by the base bias current I_{B_1}. This will raise input 2 to the dc voltage at input 1. In other words, if the currents I_{B_1} and I_{B_2} are equal, and the resistances seen by these currents are equal, the voltages to ground at inputs 1 and 2 are equal. This means that there will be no dc differential input voltage V_{id} to cause an output offset V_{oo}. The value of R_2 needed to eliminate or reduce the dc differential input voltage, V_{id}, and the resulting output offset, V_{oo}, is easily found.

Note in Fig. 4-9 that current I_{B_1} sees two parallel paths, one containing R_1 and R_s in series and the other containing R_F. Thus when $V_o \cong 0$ V, current I_{B_1} sees a resistance $R_1 + R_s$ in parallel with the feedback resistor R_F. Since current I_{B_2} is to see a resistance R_2 that is equal to the total resistance seen by I_{B_1}, we can show that

$$R_2 = \frac{(R_1 + R_s)R_F}{R_1 + R_s + R_F} \tag{4-5}$$

Example 4-4

Referring to the circuit in Fig. 4-6, what value of resistance can we use in series with the noninverting input (pin 3) to reduce or eliminate the output offset voltage caused

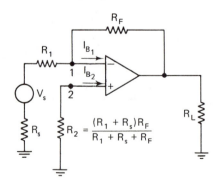

Figure 4-9 Resistor R_2 in series with the noninverting input reduces the output offset voltage caused by the input bias current I_B.

by the input bias current? Assume that the internal resistance of the signal source V_s is negligible.

Answer.   Since the input resistor $R_1 = 1\ \mathrm{k\Omega}$, the feedback resistor $R_F = 100\ \mathrm{k\Omega}$, and the signal source's resistance $R_s = 0\ \Omega$, we can use

$$R_2 = \frac{(1\ \mathrm{k\Omega} + 0)100\ \mathrm{k\Omega}}{1\ \mathrm{k\Omega} + 0 + 100\ \mathrm{k\Omega}} = 990\ \Omega$$

4.3 INPUT OFFSET CURRENT I_{io}

In the previous section we learned that a resistor R_2 placed in series with the noninverting input 2 reduces the output offset caused by the input bias current. However, Eq. (4-5) for finding the necessary value of R_2 was derived assuming that the base bias currents I_{B_1} and I_{B_2} are equal. In practice, due to imbalances within the Op Amp's circuitry, these currents are at best only approximately equal, as indicated in Eq. (4-3). The input offset current I_{io}, usually specified by the manufacturer, is a parameter that indicates how far from being equal the currents I_{B_1} and I_{B_2} can be. In fact, the input offset current is defined as the difference in the two base bias currents, that is,

$$I_{io} = |I_{B_1} - I_{B_2}| \qquad\qquad (4\text{-}6)$$

When given the value of input offset current I_{io}, we can predict how much output offset voltage a circuit like that in Fig. 4-9 might have, caused by the existence of base bias currents. Due to inequalities of the currents I_{B_1} and I_{B_2}, the voltages with respect to ground at inputs 1 and 2 will be unequal, even though a properly chosen value of R_2 is used. This causes a dc differential input voltage V_{id}, which in turn causes an output offset voltage V_{oo}. In other words, the amount of dc differential input voltage and the amount of resulting output offset depend on the amount of difference in the base bias currents I_{B_1} and I_{B_2}, which is the input offset current I_{io}. More specifically, the amount of output offset voltage V_{oo} in a circuit like that in Fig. 4-9, caused by the input offset current I_{io}, can be closely approximated with the equation*

$$V_{oo} \cong R_F I_{io} \qquad\qquad (4\text{-}7)$$

if the value of R_2 was determined with Eq. (4-5).

Example 4-5

Referring to the circuit in Fig. 4-10, find the maximum output offset caused by the input offset current I_{io}. The Op Amp is a type 741, and the internal resistance of the signal source V_s is negligible.

---

* See Appendix K for the derivation.

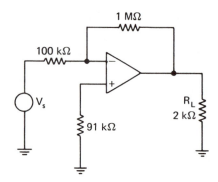

Figure 4-10

Answer.   The 741's maximum input offset current I_{io} = 200 nA as indicated on its specification sheet. Since the value of R_2 satisfies Eq. (4-5), we can closely estimate the output offset caused by I_{io} with Eq. (4-7). In this case

$$V_{oo} \cong R_F I_{io} = 1 \text{ M}\Omega(200 \text{ nA}) = 200 \text{ mV}$$

This shows that we might have an output offset due to unequal base bias currents even though a properly chosen resistor R_2 is placed between the noninverting input and ground. If R_2 is not used, however, the output offset caused by input currents is considerably larger and therefore potentially a greater problem.

Example 4-6

Referring again to the circuit in Fig. 4-10, find the maximum output offset caused by the input currents if R_2 is removed and, instead, the noninverting input is connected directly to ground. As before, the Op Amp is a type 741.

Answer.

$$V_{oo} \cong R_F I_B = 1 \text{ M}\Omega(500 \text{ nA}) = 500 \text{ mV}$$

It is interesting to note that the output offset can be more than doubled if the resistor R_2 is not used. Therefore, R_2 may or may not be necessary, depending on how important it is to have little or no output offset. The required stability of the output voltage with temperature changes must be considered too, as we will see in a later chapter. The size of the feedback resistor is also a factor. As mentioned before, if a relatively small feedback resistor is used, the effects of base bias current and the input offset current are also relatively small. Some types of IC Op Amps are made with FETs or high-beta transistors* in the input differential stage. These have dc input currents on the order of just a few nanoamperes. Some *hybrid* models of Op Amps, which contain discrete and integrated circuits, have input bias currents as low as 0.01 pA. Of course, such a small dc input bias current reduces the current-generated output offset voltage to insignificance. As we will see in applications later, an extremely small input bias current is desirable, and in fact necessary, in long-term integrating and sample-and-hold circuits.

---

* Op Amps with high-beta input transistors are sometimes called *super beta* transistor input Op Amps. An extremely high beta can be achieved by connecting conventional transistors in a Darlington pair.

4.4 COMBINED EFFECTS OF V_{io} AND I_{io}

In a circuit such as that in Fig. 4-9, the input offset voltage V_{io} can cause either a positive or a negative output offset voltage V_{oo} [see Eq. (4-1)]. Likewise, the input offset current I_{io} can cause either a positive or a negative output offset voltage V_{oo} [see Eq. (4-7)]. The effects of input offset voltage V_{io} and input offset current I_{io} might buck and cancel each other, resulting in little output offset. On the other hand, they might be additive, causing an output offset voltage V_{oo} that is the sum of the output offsets caused by V_{io} and I_{io} working independently. Thus the total output offset voltage in the circuit of Fig. 4-9 can be as large as

$$V_{oo} \cong A_v V_{io} + R_F I_{io} \qquad (4\text{-}8\text{a})$$

if the value of R_2 satisfies Eq. (4-5). This equation is sometimes shown in other equivalent forms, such as

$$V_{oo} \cong [V_{io} + R_2 I_{io}]\left(\frac{R_F}{R_1} + 1\right) \qquad (4\text{-}8\text{b})$$

or

$$V_{oo} \cong [V_{io} + R_2(I_{B_1} - I_{B_2})]\left(\frac{R_F}{R_1} + 1\right) \qquad (4\text{-}8\text{c})$$

Example 4-7

Considering the effects of both V_{io} and I_{io}, what is the maximum output offset voltage V_{oo} of the circuit in Fig. 4-10 if the Op Amp is a type 741?

Answer.   Arbitrarily selecting Eq. (4-8b) above, we can show that the maximum output offset voltage is

$$V_{oo} \cong [6 \text{ mV} + 91 \text{ k}\Omega(200 \text{ nA})]\left(\frac{1000 + 100}{100}\right) = 266 \text{ mV}$$

REVIEW QUESTIONS

4-1. What is the input offset voltage of an Op Amp?

4-2. Why is an input offset voltage V_{io} often needed to null the output of an Op Amp?

4-3. What polarity of required input offset voltage would you expect at input 1 with respect to input 2?

4-4. What is the meaning of the term *base bias current*?

4-5. What are the difference and similarity of the base bias and input bias currents?

4-6. What problem might exist if a significant base bias current exists and if the feedback resistor is relatively large?

4-7. How can the undesirable effect of a significant input bias current be minimized?

4-8. The nulling circuits in Figs. 4-2 and 4-3 are able to provide a dc differential
input voltage that is (*positive*), (*negative*), (*either polarity*) at input 1 with
respect to input 2.

4-9. The nulling circuits in Figs. 4-2 and 4-3 are able to adjust the output V_o to zero
volts in spite of the existence of (*input offset voltage*), (*input bias current*),
(*input offset voltage and input bias current*).

4-10. What is the meaning of the term *input offset current*?

PROBLEMS

4-1. If the Op Amp in the circuit of Fig. 4-11 is a type 741, what is the circuit's
closed-loop voltage gain if the potentiometer (POT) is adjusted to 39 kΩ? The
8-pin DIP (dual-in-line) package has the same pin identifications as does the 8-
pin circular metal can package.

4-2. If the Op Amp in the circuit of Fig. 4-11 is a LF13741 type (Appendix D), what
is the circuit's closed-loop gain if the POT is adjusted to 14 kΩ? It has the same
pin identifications as does the 741 8-pin circular metal can package or DIP.

4-3. If the slide on the POT is moved to the far right in the circuit of Fig. 4-11, what
is the approximate closed-loop voltage gain of the circuit?

4-4. If the slide on the POT is moved to the far left in the circuit of Fig. 4-11, what is
the approximate closed-loop voltage gain of the circuit?

4-5. In the circuit described in Prob. 4-1, what is the maximum output offset caused
by the Op Amp's input offset voltage?

4-6. What is the maximum output offset caused by the input offset voltage in the
circuit described in Prob. 4-2?

4-7. In Prob. 4-1, what is the circuit's maximum output offset caused by the
existence of base bias currents in the input leads?

4-8. What is the maximum possible output offset voltage caused by the existence of
input bias currents in the circuit described in Prob. 4-2?

4-9. Referring to the circuit described in Prob. 4-1, what maximum output offset
voltage might it have with the combined effects of input offset voltage and input
bias currents?

4-10. Referring to the circuit described in Prob. 4-2, what maximum output offset
voltage might it have with the combined effects of input offset voltage and input
bias currents?

4-11. If the POT in the circuit of Fig. 4-11 is replaced with a fixed 100-kΩ resistor,
what value of resistance should we put in series with pin 3 and ground to
reduce the effect of input bias currents? Assume that the Op Amp is a type 741
and that the internal resistance of the signal source is negligible.

4-12. If the POT in the circuit of Fig. 4-11 is replaced with a fixed 10-kΩ resistor,
what value of resistance should we put in series with pin 3 and ground to

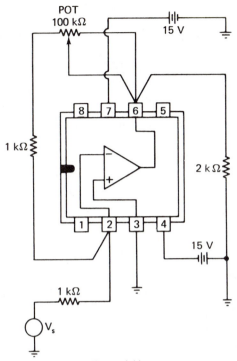

Figure 4-11

reduce the effect of input bias current? The Op Amp is a type LF13741 and the internal resistance of the signal source is negligible.

4-13. In the circuit described in Prob. 4-11, what is the maximum output offset voltage caused by the existence of input bias currents if the properly selected resistance is used between pin 3 and ground?

4-14. In the circuit described in Prob. 4-12, what is the maximum output offset caused by the existence of input bias currents if the properly selected resistance is used between pin 3 and ground?

4-15. In the circuit of Fig. 4-12, if full-range adjustment of the 50-kΩ POT almost, but not quite, nulls the output, what small modification can we make to have full null control?

4-16. If in the circuit of Fig. 4-12 the input signal V_s is a sine wave but the output has the waveform shown in Fig. 4-13, what modification or adjustment will probably correct the problem?

4-17. If the circuit in Fig. 4-12 has a maximum peak-to-peak output capability of 20 V, what maximum peak-to-peak input signal V_s can we apply?

4-18. If the circuit in Fig. 4-12 is nulled, has a maximum peak-to-peak output capability of 20 V, and has an input sine-wave voltage V_s of 2 V peak-to-peak, what modification is necessary in this circuit if clipping of the output signal is to be avoided?

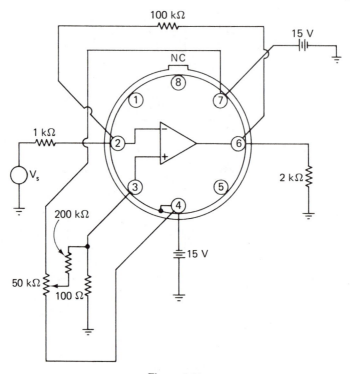

Figure 4-12

4-19. If V_s is a sine wave in the circuit of Fig. 4-14, and if an oscilloscope reading shows a sine-wave voltage at pin 6 to ground with a peak-to-peak value of 6.3 V, what is the peak-to-peak amplitude of the input V_s?

4-20. If the circuit in Fig. 4-14 has the waveform shown in Fig. 4-13 at pin 6 with respect to ground, what is the approximate peak-to-peak amplitude of V_s?

4-21. What is the purpose of the 10-kΩ POT in the circuit of Fig. 4-14 if the Op Amp is a type 741?

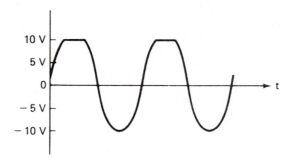

Figure 4-13

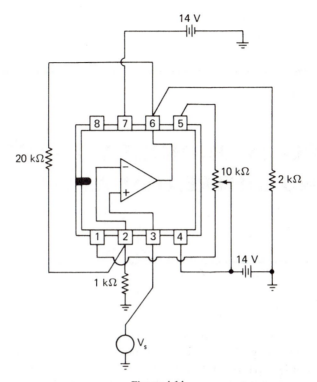

Figure 4-14

4-22. If the Op Amp in the circuit of Fig. 4-14 is a type 741, what is its maximum peak-to-peak output signal capability?

4-23. If the Op Amp in the circuit of Fig. 4-15 is a type 741, what is its maximum possible output offset, caused by both the input offset voltage and the input offset current?

4-24. If the Op Amp in the circuit of Fig. 4-15 is a type LF13741, what is its maximum possible output offset, caused by both the input offset voltage and the input offset current?

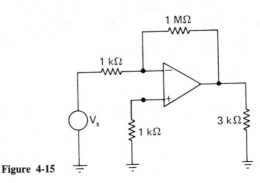

Figure 4-15

5

COMMON-MODE VOLTAGES
AND DIFFERENTIAL AMPLIFIERS

A voltage appearing at both inputs of an Op Amp, as shown in Fig. 5-1, is a common-mode voltage V_{cm}. A common-mode voltage V_{cm} can be dc, ac, or a combination (superposition) of dc and ac. Thus a common-mode voltage V_{cm} might be an unavoidable dc level on both inputs. When Op Amps are required to work in time-varying magnetic and electric fields, ac voltage can be induced into both input leads. In such cases, dc and ac common-mode voltages V_{cm} can exist simultaneously at an Op Amp's inputs. The induced ac voltage includes 60 Hz from nearby electrical power equipment. Induced voltages can be higher frequencies or even irregular transients. In any case, it is undesired and considered as *noise* in the system.

Amplifiers with differential inputs (e.g., Op Amps) have more or less ability to reject common-mode voltages. This means that, although fairly large common-mode voltages might exist at an Op Amp's differential inputs, these voltages can be reduced to very small and often insignificant amplitudes at the output. In this chapter we will become familiar with manufacturers' specifications that will enable us to predict how well a given Op Amp will reject (not pass) common-mode voltages. We will also see applications in which the Op Amp's ability to reject common-mode voltages is useful.

5.1 THE DIFFERENTIAL-MODE OP AMP CIRCUIT

To appreciate fully how a properly wired Op Amp circuit is able to reject undesirable common-mode voltages, we should first look at the familiar circuits in which induced noise is potentially a problem. For example, for

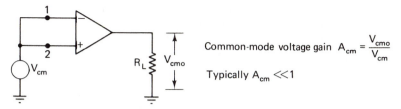

Common-mode voltage gain $A_{cm} = \dfrac{V_{cmo}}{V_{cm}}$

Typically $A_{cm} \ll 1$

Figure 5-1   Op Amp with common-mode voltage applied.

both circuits shown in Fig. 5-2, voltage V_s is the desired signal to be amplified and is typically from a preceding transducer or amplifier. If the input lead has any appreciable length, and if varying fields are present, noise voltage V_n will probably be induced into it as shown. In both of these circuits, the Op Amp cannot distinguish noise V_n from desired signal V_s; both are amplified and appear at the output. Thus if the closed-loop gain A_v of each of these circuits is about 100, the output noise voltage V_{no} and output signal voltage V_o are about 100 times larger than their respective inputs.

Noise voltages at the output of an Op Amp are greatly reduced if the circuit is connected to operate in a differential mode, as shown in Fig. 5-3a.

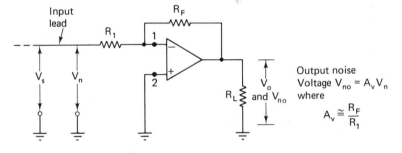

Output noise
Voltage $V_{no} = A_v V_n$
where

$$A_v \cong \frac{R_F}{R_1}$$

(a) Inverting-mode amplifier

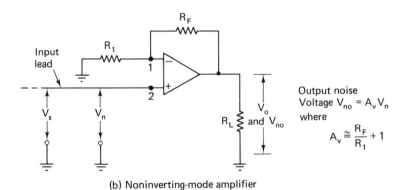

Output noise
Voltage $V_{no} = A_v V_n$
where

$$A_v \cong \frac{R_F}{R_1} + 1$$

(b) Noninverting-mode amplifier

Figure 5-2   Inverting-mode and noninverting-mode circuits amplify induced noise.

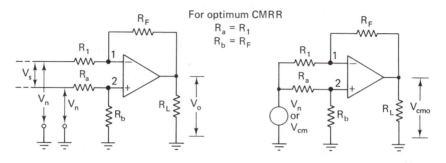

(a) Op Amp connected in differential mode; $A_V \cong -\dfrac{R_F}{R_1}$.

(b) Equivalent of differential-mode circuit showing induced noise voltage V_n as a common-mode voltage V_{cm}.

Figure 5-3

In this circuit, the desired signal V_s is amplified normally because it is applied *across* the two inputs. That is, signal V_s causes a differential input V_{id} to appear across the input terminals 1 and 2. The noise voltage V_n, however, is induced into each input lead with respect to ground or a common. With proper component selection, both of the induced noise voltages V_n are equal in amplitude and phase and therefore are common-mode voltages as shown in Fig. 5-3b. Thus if the noise voltage to ground at input 1 is the same as the noise voltage to ground at input 2, the differential noise voltage across input terminals 1 and 2 is zero. Since ideally an Op Amp amplifies only differential input voltages no noise voltage should appear at the output. However, due to inherent electrical characteristics of an Op Amp—such as internal capacitances—some common-mode signals will get through to the load R_L. The ratio of the output common-mode voltage V_{cmo} to the input common-mode voltage V_{cm} is called the common-mode gain A_{cm}, that is,

$$A_{cm} = \frac{V_{cmo}}{V_{cm}} \tag{5-1}$$

Ideally, this gain A_{cm} is zero. In practice, it is finite but usually much smaller than 1.

5.2 COMMON-MODE REJECTION RATIO, *CMRR*

Op Amp manufacturers do not list a common-mode gain factor. Instead, they list a *common-mode rejection ratio*, CMRR. The CMRR is defined in several essentially equivalent ways by the various manufacturers. It can be defined as the ratio of the change in the input common-mode voltage V_{cm} to the resulting change in input offset voltage V_{io}. Thus

$$CMRR = \frac{\Delta V_{cm}}{\Delta V_{io}} \tag{5-2}$$

It can also be shown to be approximately equal to the ratio of the closed-loop gain A_v to the common-mode gain A_{cm}, that is,

$$CMRR = \frac{A_v}{A_{cm}} \tag{5-3}$$

Generally, larger values of *CMRR* mean better rejection of common-mode signals and are therefore more desirable in applications where induced noise is a problem and the differential-mode circuit is used. Later, we will see that the *CMRR* of an Op Amp tends to decrease with higher common-mode frequencies.

The common-mode rejection is usually specified in decibels (dB), where

$$CMR(\text{dB}) = 20 \log \frac{\Delta V_{cm}}{\Delta V_{io}} \tag{5-4a}$$

or

$$CMR(\text{dB}) = 20 \log \frac{A_v}{A_{cm}} = 20 \log CMRR \tag{5-4b}$$

A chart for converting common-mode rejection from a ratio to dBs, or vice versa, is given in Fig. 5-4.

Equation (5-3) shows that the common-mode voltage gain A_{cm} of a differential-mode Op Amp circuit is a function of its closed-loop gain and the Op Amp's specified *CMRR*. Rearranging Eq. (5-3), we can show that

$$A_{cm} = \frac{A_v}{CMRR} \tag{5-3}$$

The equivalent of this in dBs can be shown as

$$20 \log A_{cm} = 20 \log A_v - 20 \log CMRR \tag{5-5a}$$

or as

$$A_{cm}(\text{dB}) = A_v(\text{dB}) - CMR(\text{dB}) \tag{5-5b}$$

Since the closed-loop gain A_v is usually much smaller than the *CMRR*, the common-mode gain A_{cm} in Eq. (5-3) is much smaller than 1. In either version of Eq. (5-5) then, the common-mode gain in dBs is negative.

Some manufacturers define *CMRR* as the reciprocal of the right side of Eq. (5-2). This implies that the *CMRR* in dB is a negative value. If we assume that the specified value of *CMR*(dB) is positive, then we subtract it

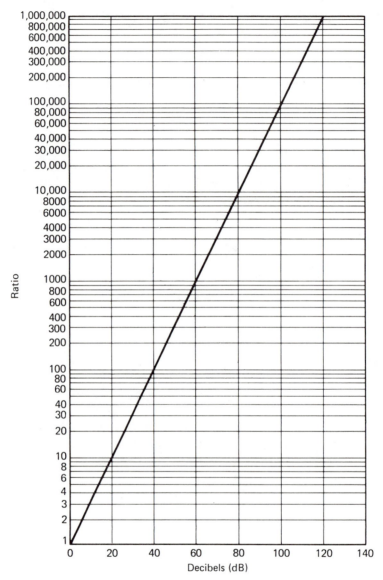

Figure 5-4 Chart for converting voltages ratios to dB and vice versa.

from $A_v(\text{dB})$ to obtain the value of $A_{cm}(\text{dB})$, as shown in Eq. (5-5b). If we assume that $CMR(\text{dB})$ is negative, then we add it to $A_v(\text{dB})$ to get $A_{cm}(\text{dB})$. In either case, we solve for the *difference* in the values on the right side of Eq. (5-5) to determine $A_{cm}(\text{dB})$.

The signal voltage gain V_o/V_s of the differential circuit in Fig. 5-3a is determined with Eq. (3-4), that is,

$$A_v = \frac{V_o}{V_s} \cong - \frac{R_F}{R_1} \tag{3-4}$$

As shown in Fig. 5-3a, the resistors R_a and R_b are selected to be equal to R_1 and R_F, respectively. This assures us of equal resistances to ground, looking into either input lead. It also provides equal induced noise voltages to ground at input terminals 1 and 2 and thus makes noise voltages common mode, enabling the Op Amp to reject the induced noise, more or less depending on its *CMRR* value. Resistors R_a and R_b also serve the same function as R_2 does in the circuit of Fig. 4-9—i.e., to reduce the output offset V_{oo} caused by input bias current I_B.

Example 5-1

Referring to the circuit in Fig. 5-3a, if $R_1 = R_a = 1 \text{ k}\Omega$, $R_F = R_b = 10 \text{ k}\Omega$, $V_s = 10 \text{ mV}$ at 1000 Hz, and $V_n = 10 \text{ mV}$ at 60 Hz, what are the amplitudes of the 1000-Hz signal and the 60-Hz noise at the load R_L? The Op Amp's $CMR(\text{dB}) = 80$ dB.

Answer.   This circuit's closed-loop gain is

$$A_v \cong - \frac{R_F}{R_1} = - \frac{10 \text{ k}\Omega}{1 \text{ k}\Omega} = - 10$$

The negative sign means that the signal voltage output with respect to ground is out of phase with the signal at the inverting input 1 with respect to the noninverting input 2. Since the 1000 Hz is applied in a differential mode, it is amplified by this gain factor. Thus, at 1000 Hz, the output signal is

$$V_o \cong -10(10 \text{ mV}) = -100 \text{ mV}$$

The 60-Hz noise voltage, on the other hand, is applied in common mode. Therefore, we can first find the common-mode gain with Eq. (5-3), that is,

$$A_{cm} = \frac{A_v}{CMRR} \cong \frac{-10}{10,000} = -1 \times 10^{-3}$$

where 80 dB is equal to 10,000 according to Fig. 5-4.

Since the 60-Hz noise is the common-mode voltage, we can find the amount of output common-mode voltage with Eq. (5-1), that is,

$$V_{cmo} = A_{cm}V_{cm} \cong -1 \times 10^{-3}(10 \text{ mV}) = -10\mu\text{V}$$

Thus, the 60-Hz output is smaller than the input induced 60 Hz, by the factor 1000, and the Op Amp is shown as an effective tool in reducing noise problems. The negative sign of V_{cmo} here is of no significance. V_{cmo} can be in phase with V_{cm} as often as out of phase.

We could have determined the common-mode output V_{cmo} using Eq. (5-5). For example, since a closed-loop gain of 10 is equivalent to 20 dB according to Fig. 5-4, then by Eq. (5-5b),

$$A_{cm}(\text{dB}) = 20 - 80 = -60 \text{ dB}$$

The negative sign means that the common-mode voltage is attenuated by 60 dB, which means that V_{cmo} is smaller than V_{cm} by the factor 1000. Note that 60 dB is equivalent to the ratio 1000 in Fig. 5-4. Thus $V_{cmo} = V_{cm}/1000$, etc.

5.3 MAXIMUM COMMON-MODE INPUT VOLTAGES

A common-mode input voltage can be dc as in the circuit of Fig. 5-5. In this case, the bridge circuit has one dc supply voltage E. By the voltage divider action of resistances R, a portion of this supply E is applied to points I and II. The voltages at these points are shown as V_I and V_{II}. Their average value is the common-mode input, that is,

$$V_{cm} = \frac{V_I + V_{II}}{2} \tag{5-6}$$

Further voltage divider action takes place by resistors R_a and R_b and by resistors R_1 and R_F—i.e., the actual voltage applied to the noninverting input 2 is

$$V_2 = \frac{V_{II}R_b}{R_a + R_b} \tag{5-7}$$

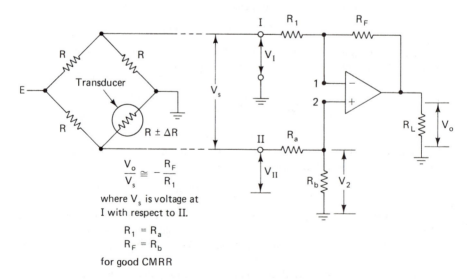

Figure 5-5 Differential-mode circuit with a dc common-mode input.

Since V_1 is *virtually* at the same potential as V_2, then we can similarly show that

$$V_1 \cong \frac{V_{II}R_b}{R_a + R_b} \tag{5-8}$$

The voltage V_1 or V_2 should never exceed the maximum *input voltage* or the maximum *common-mode voltage* specified on an Op Amp's data sheets.

5.4 OP AMP INSTRUMENTATION CIRCUITS

Frequently, in instrumentation and industrial applications, the Op Amp is used to amplify signal output voltages from bridge circuits. We considered such a circuit earlier in Fig. 3-17a. Similar circuits are the useful *half-bridge* circuits as shown in Fig. 5-6. Because these circuits are single-ended—i.e., each of their Op Amps is connected to work in an inverting mode or a noninverting mode—they are prone to induced noise problems. Therefore, if the bridge-type transducer circuit is required to work in time-varying electric and magnetic fields, its associated amplifier should be a differential type, such as in Fig. 5-5.

In the circuit of Fig. 5-5, the maximum specified common-mode input voltage of the Op Amp dictates the maximum allowable dc source voltage E on the bridge. The transducer's resistance changes by ΔR when the appropriate change in its physical environment occurs. The bridge converts this physical change to a voltage change across points I and II which is amplified by the Op Amp. Thus, depending on the type of transducer used, a change in temperature, light intensity, strain, or whatever, is converted to an amplified electrical signal. The gain of the differential amplifier can be increased or decreased if we increase or decrease *both* R_F and R_b by equal amounts. In other words, R_F and R_b must remain equal if good *CMR* is to be retained. However, if R_F and R_b are changed, the resistances looking into points I and II will change noticeably too. The accompanying problems with this kind of variable loading on the bridge can be avoided by use of high input impedance circuits such as in Fig. 5-7. The gain factors of these circuits are shown to be negative, representing the out-of-phase relationship of V_o and V_s, where V_s is measured from input I with respect to input II.

The circuit shown in Fig. 5-7a is simply a differential-mode amplifier preceded by a voltage follower at each input. These voltage followers (buffers) have extremely large input impedances (resistances) and therefore draw negligible current from the source of signal V_s. Thus, even if the gain of the differential stage is changed, the resistance into inputs I and II remains large and constant. Since voltage followers have unity gain, the gain of the

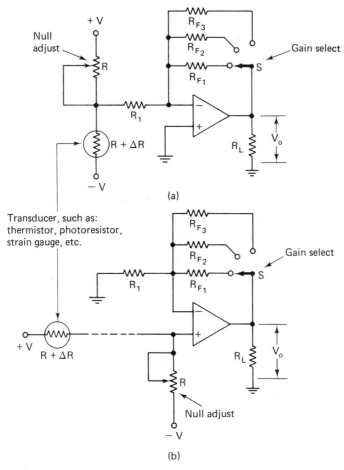

Figure 5-6   Half-bridge working into (a) an inverting circuit and (b) a noninverting circuit.

entire circuit is the gain of the output differential stage. Thus, for the circuit shown in Fig. 5-9a,

$$\frac{V_o}{V_s} \cong -\frac{R_F}{R_1} \tag{3-4}$$

The circuit in Fig. 5-7b also has high input impedance due to a noninverting-mode amplifier at each input. This circuit's differential gain V_o/V_s is determined by the components connected on the output (lower) Op Amp; that is

$$\frac{V_o}{V_s} \cong -\left(1 + \frac{R'_F}{R'_1}\right) \tag{5-9}$$

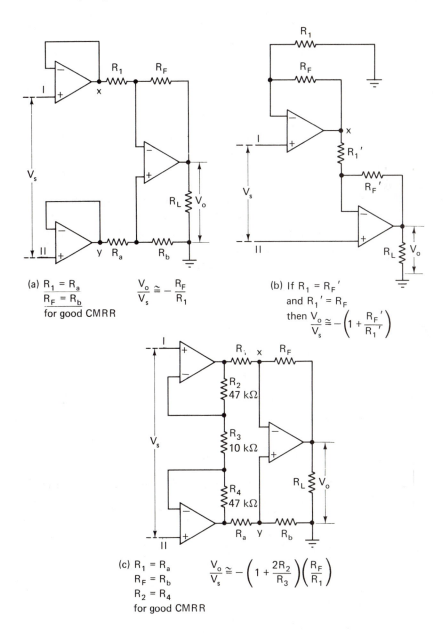

(a) $R_1 = R_a$
$R_F = R_b$
for good CMRR

$$\frac{V_o}{V_s} \cong -\frac{R_F}{R_1}$$

(b) If $R_1 = R_F'$
and $R_1' = R_F$
then $\frac{V_o}{V_s} \cong -\left(1 + \frac{R_F'}{R_1'}\right)$

(c) $R_1 = R_a$
$R_F = R_b$
$R_2 = R_4$
for good CMRR

$$\frac{V_o}{V_s} \cong -\left(1 + \frac{2R_2}{R_3}\right)\left(\frac{R_F}{R_1}\right)$$

Figure 5-7 High input impedance differential-mode amplifiers. The negative signs in the gain equations mean phase reversal if V_s is measured from the upper input with respect to the lower input.

Once the gain and components R'_1 and R'_F are selected, the components in the upper stage must comply with the equations

$$R_F = R'_1$$

and

$$R_1 = R'_F$$

A useful instrumentation amplifier is shown in Fig. 5-7c. It has high input resistance, a good *CMRR*, and variable gain that is adjustable with the potentiometer R. Generally, smaller values of R provide larger gains. The equation for gain is easy to derive. Note in Fig. 5-8 that because $V_{id} \cong 0$ on both buffers, the input signal V_s must appear across the potentiometer R. Also we note that the output signal V_{o_1} of these buffers is across both resistors R_2 and the potentiometer R. These carry the same current and can be viewed as series resistors. By Ohm's law, then,

$$\frac{V_{o_1}}{V_s} = \frac{(R_2 + R + R_2)I}{RI} = \frac{(2R_2 + R)I}{RI}$$

where I is the signal current which can be cancelled to show that these buffers have a gain of

$$\frac{V_{o_1}}{V_s} = \frac{(2R_2 + R)}{R} = \frac{2R_2}{R} + 1$$

The differential stage, following the buffers, has a gain of $-R_F/R_1$. This, combined with the gain of the buffers, yields a total gain of

$$\frac{V_o}{V_s} \cong -\left(1 + \frac{2R_2}{R}\right)\left(\frac{R_F}{R_1}\right) \tag{5-10}$$

Another variable-gain differential-mode amplifier is shown in Fig. 5-9. Its gain V_o/V_s is varied by an adjustment of the potentiometer R_3, though its

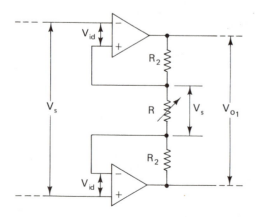

Figure 5-8 Buffer stage for instrumentation amplifier has variable gain.

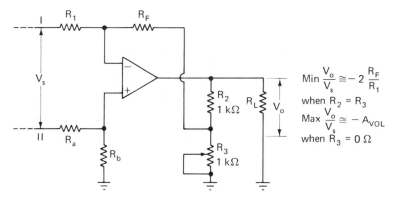

Figure 5-9 Differential amplifier with variable gain.

CMRR is affected thereby. Generally, the gain is increased or decreased by a decrease or increase, respectively, of R_3. In this circuit, the right end of the feedback resistor R_F is terminated with a voltage divider and the output signal voltage V_o is across this divider. When R_3 is maximum, half as much signal is fed back to the inverting input via R_F than would be the case if R_F were connected directly to the output as in the simpler differential amplifier in Fig. 5-3a. Thus the minimum gain can be estimated with the equation:

$$\frac{V_o}{V_s} \cong -2 \left(\frac{R_F}{R_1} \right) \qquad (5\text{-}11)$$

This shows that twice as much gain occurs with half as much negative feedback. If R_3 is adjusted to a minimum of 0 Ω, the right end of R_F is grounded and no negative feedback occurs. This tends to pull the stage gain up to the open-loop gain A_{VOL} of the Op Amp. Of course, both inputs can be preceded with voltage followers if high input impedance is required.

REVIEW QUESTIONS

5-1. What is a common-mode voltage?

5-2. What is the ideal value of *CMRR*?

5-3. If an Op Amp is to work in an environment that has significant induced voltages, does grounding *one* of the inputs reduce noise in the output? Why?

5-4. What do the initials *CMRR* mean?

5-5. Graphically, convert the following *CMRR* values to decibels: 10, 100, 1000, 10,000, 100,000, and 10^6.

5-6. What is the ideal value of common-mode voltage gain?

5-7. In what configuration is the Op Amp arranged to make induced noise voltages common mode?

5-8. What effect do the dc supply voltage values have on the maximum recommended common-mode voltages?

5-9. Assume that in the circuit of Fig. 5-3a, R_F is replaced with a potentiometer, while all other resistors remain fixed. What will happen to the circuit's differential gain A_v and its *CMRR* capability if the potentiometer is varied?

5-10. What is the purpose of using a voltage follower in each input of a differential amplifier?

PROBLEMS

5-1. In the circuit of Fig. 5-10, the output V_{cmo} measures 4 mV rms, 60 Hz, when the applied V_{cm} = 400 mV rms, 60 Hz. What is the Op Amp's common-mode rejection in decibels.

5-2. In the circuit of Fig. 5-10, the output V_{cmo} measures about 127 μ V rms, 60 Hz, when the applied V_{cm} = 400 mV rms, 60 Hz. What is this Op Amp's common-mode rejection in decibels?

5-3. In the circuit of Fig. 5-11, if the signal voltage at input I varies from -30 mV up to 20 mV and if 10 mV rms, 60 Hz, is present at input I to ground, what are (a) the signal variations at output 0 and (b) the 60-Hz rms voltage V_n at output 0? The Op Amp's *CMRR* = 10,000.

5-4. If the signal at input I to ground in the circuit of Fig. 5-11 varies from -25 mV to $+ 35$ mV while 15 mV rms, 60 Hz is also present at this input, what are (a) the signal variations and (b) the 60-Hz rms voltage V_n at output 0? The Op Amps's *CMRR* = 15,000.

5-5. Refer to the circuit in Fig. 5-12. If $R_F = R_b = 10$ kΩ, $R_1 = R_a = 1$ kΩ, *CMR*(dB) = 80 dB, signal across inputs I and II varies between -0.5 V and $+ 0.5$V, and if both of these inputs have $+2$ V dc and 20 mV rms, 60 Hz, potentials with respect to ground, find (a) the approximate dc voltages to ground at inputs 1 and 2, (b) the output signal variations, and (c) the rms value of 60-Hz voltage at output 0.

5-6. If in the circuit of Fig. 5-12 $R_F = R_b = 20$ kΩ, $R_1 = R_a = 500$ Ω, *CMRR* = 50,000, signal across inputs I and II varies from -80 mV to $+80$ mV, and if both of these inputs have $+4$ V dc and 600 mV rms, 60 Hz, potentials to

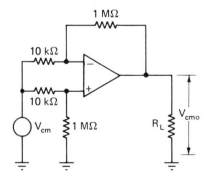

Figure 5-10

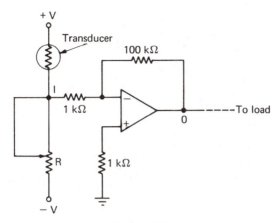

Figure 5-11

ground, find (a) the approximate dc voltages to ground at inputs 1 and 2, (b) the output signal variations, and (c) the rms value of the 60-Hz voltage at output 0.

5-7. Suppose that in the circuit of Fig. 5-12, the supply voltages on the bridge and Op Amps are ± 15 V, $R_F = R_b = 10$ kΩ, $R_1 = R_a = 1$ kΩ and $R_y = 4$ kΩ. The Op Amp's output becomes 0 V (nulled) when R_z is adjusted to 1 kΩ. Find: (a) the dc common-mode input voltage, and (b) the dc to ground voltage at inputs 1 and 2 of the Op Amp.

5-8. Referring to the circuit described in the previous problem and to the Commercial Selection Guide of Appendix D, if the Op Amp is a LM741C, has its maximum common-mode input voltage been exceeded? Explain.

5-9. In the circuit of Fig. 5-12, $R_x = R_y = 1$ kΩ and a null is obtained when R_z is adjusted to 1 kΩ. The bridge and Op Amp supply voltages are ± 15 V, $R_1 = R_a = 110$ kΩ and $R_F = R_b = 220$ kΩ. What is (a) the dc common-mode input voltage and (b) the range of output voltage V_o if the transducer's resistance varies over the range 900 Ω to 1.1 kΩ?

5-10. In the circuit of Fig. 5-13, $R_y = 3.3$ kΩ, $R_z = 2.2$ kΩ, and a null is obtained when R_x is adjusted to 300 Ω. The bridge supply voltage $E = 5$ V and the Op

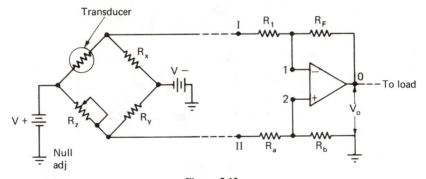

Figure 5-12

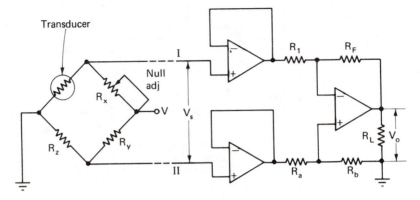

Figure 5-13

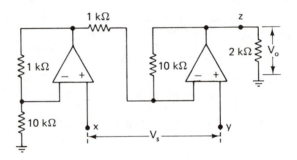

Figure 5-14

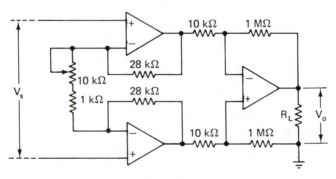

Figure 5-15

Amp's supply voltages are ± 15 V. If $R_1 = R_a = 1.1$ kΩ and $R_F = R_b = 3.3$ kΩ, what are (a) the dc common-mode input voltage and (b) the range of output voltage V_o if the transducer's resistance varies over the range 190 Ω to 210 Ω?

5-11. Referring to the circuit described in Prob. 5-9, over what range will the voltage at point I with respect to point II vary if the transducer's resistance varies from 0 Ω to infinity?

5-12. Referring to the circuit described in Prob. 5-10, over what range will the voltage V_s vary if the transducer's resistance varies from 0 Ω to infinity?

5-13. In Prob. 5-9, if V_o clips at ± 12 V, what maximum ratios of R_F/R_1 and R_b/R_a can we use and still avoid such clipping?

5-14. In Prob. 5-10, if V_o clips at ± 12 V, what maximum ratios of R_F/R_1 and R_b/R_a can we use and still avoid such clipping?

5-15. What is the gain V_o/V_s of the circuit in Fig. 5-14?

5-16. What is the gain V_o/V_s of the circuit in Fig. 5-14 if both 1-kΩ resistors are replaced with 2.2 kΩ values?

5-17. In the circuit of Fig. 5-14, if the voltage at point x is $+5.5$ V to ground and the voltage at y is $+5.1$ V to ground, what is the voltage at point z with respect to ground, assuming that the output was initially nulled? What is the average dc common-mode input voltage?

5-18. In the circuit of Fig. 5-14, if the voltages at points x and y are -0.1 V and -0.5 V, respectively, what is the voltage at point z with respect to ground, and what is the dc common-mode input voltage? Assume that the output was initially nulled.

5-19. Over what range can the gain V_o/V_s in the circuit of Fig. 5-15 be adjusted?

5-20. If the 1-MΩ resistors and the 28-kΩ resistors are replaced with 10-kΩ values in the circuit of Fig. 5-15, over what range can its gain be adjusted?

5-21. What is the gain V_o/V_s of the circuit in Fig. 5-15 if its 1-kΩ resistor becomes open?

5-22. In the circuit of Fig. 5-15, if the 1-kΩ resistor is replaced with a short, to what theoretical maximum can this circuit's voltage gain be adjusted?

5-23. In the circuit of Fig. 5-9, $R_1 = R_a = 10$ kΩ, and $R_F = R_b = 22$ kΩ. Find its gain V_o/V_s when (a) R_3 is adjusted to 0 Ω and when (b) R_3 is adjusted to 1 kΩ

5-24. Referring to the circuit and components described in the previous problem, if inadvertently the fixed 1 kΩ resistor and 1 kΩ potentiometer are wired in each other's places, what is V_o/V_s when (a) the potentiometer resistance is 0 Ω and when (b) its resistance is 1 kΩ?

The remaining questions refer to the circuit of Fig. 5-16. The resistors R_1 in the LH0036 have no effect on the differential gain V_o/V_s of this circuit.

5-25. What are the purposes of the potentiometers R_a and R_4?

5-26. If $R_a = 0$ Ω, $R_b = 200$ Ω, and $R_2 = 25$ kΩ, what is the gain of this circuit?

5-27. If $R_a = 20$ kΩ, $R_b = 200$ Ω, and $R_2 = 25$ kΩ, what is the gain of this circuit?

5-28. If R_a is replaced with an open, what is the gain of this circuit?

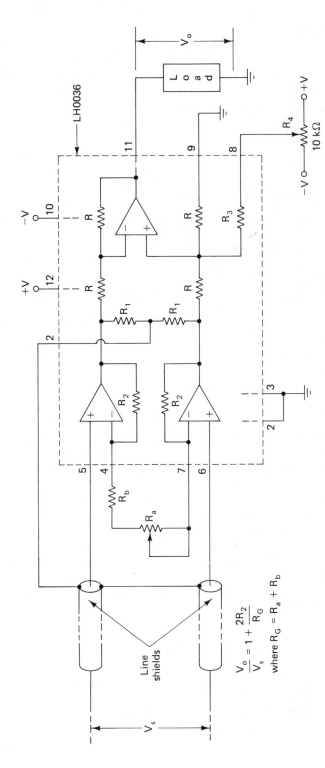

Figure 5-16 Typical application of an instrumentation amplifier LH0036.

$$\frac{V_o}{V_s} = 1 + \frac{2R_2}{R_G}$$

where $R_G = R_a + R_b$

THE OP AMP'S BEHAVIOR
AT HIGHER FREQUENCIES

Some of the characteristics of the practical Op Amp are sensitive to changes in operating frequency. In this chapter, we will consider problems caused by the decrease (roll-off) of the open-loop gain at higher frequencies. We will also see how the gain vs. frequency curve of an Op Amp can be altered, either by the circuit designer who uses the Op Amp or by the manufacturer. Other notable frequency-related problems such as reduced output-voltage swing capabilities, limited slew rates, and noise are also described here.

6.1 GAIN AND PHASE SHIFT VS. FREQUENCY

Ideally, an Op Amp should have an infinite bandwidth. This means that, if its open-loop gain is 90 dB with dc signals, its gain should remain 90 dB through audio and on to high radio frequencies. The practical Op Amp's gain, however, decreases (rolls off) at higher frequencies as shown in Fig. 6-1. This gain roll-off is caused by capacitances within the Op Amp circuitry. The reactances of these capacitances decrease at higher frequencies, causing shunt signal current paths, and thus reducing the amount of signal available at the output terminal. Along with decreased gain at higher frequencies, there is an increased phase shift of the output signal with respect to the input (see Fig. 6-2). Normally at low frequencies, there is a 180° phase difference in the signals at the inverting input and output terminals. But at higher frequencies, the output signal lags by more than 180°, and this is called *phase shift*. Thus the Op Amp with the characteristics in Fig. 6-2 has practically no

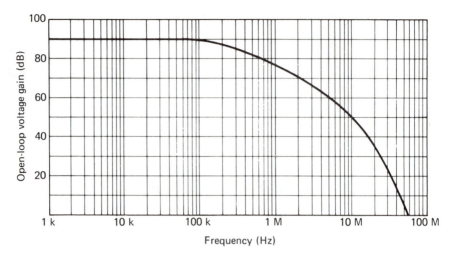

Figure 6-1 Typical open-loop gain vs. frequency curve of an Op Amp.

phase shift up to about 30 kHz. Beyond 30 kHz, the output signal starts to lag, and at 300 kHz it lags an additional 40°. This negative or lagging shift adds to the initially lagging 180°, causing an output signal lag of 220° compared with the signal applied to the inverting input. Similarly, at 1 MHz,

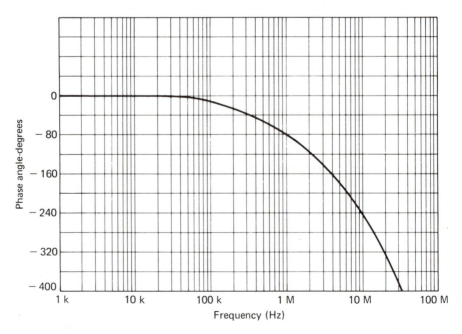

Figure 6-2 Typical open-loop output-signal phase shift vs. frequency.

the output lags the inverting input's signal by an additional 80°, or a 260° total. A serious problem caused by this kind of output-signal phase shift is discussed later in this chapter.

6.2 BODE DIAGRAMS

The gain and phase shift curves are often approximated with a series of straight lines.* Such straight-line approximations of gain and phase shift vs. frequency, where frequency is plotted on a logarithmic scale, are called *Bode diagrams* or *Bode plots*. A straight-line approximation of the curve in Fig. 6-1 is shown in Fig. 6-3. Note that, according to the approximated curve, this Op Amp's frequency response is flat from low frequencies (including dc) to 200 kHz. This means that its gain is constant from zero to 200 kHz, and therefore the bandwidth BW is about 200 kHz. Note that at higher frequencies, from 200 kHz to 2MHz, the gain drops from 90 dB to 70 dB, which is at a −20 dB/decade† or −6 dB/octave rate. The negative sign

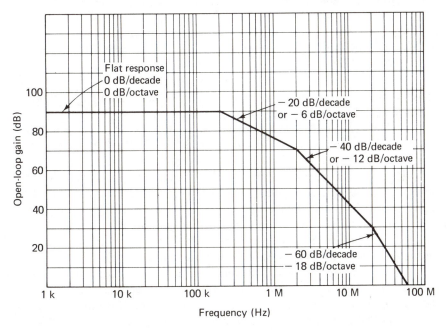

Figure 6-3.   Approximation of open-loop gain vs. frequency curve of Fig. 6-1.

* The straight lines are linear sums of the asymptotes on the individual stage gain vs. frequency curves.

† A −20 dB/decade gain roll-off means that the open-loop gain decreases 20 dB with a frequency increase by the factor 10, and is equivalent to a −6 dB/octave roll-off. A frequency change by an octave is a change by the factor 2.

refers to the negative (decreasing) slope of the curve. At frequencies from 2 MHz to 20 MHz, the roll-off is −40 dB/decade or −12 dB/octave. Above 20 MHz, the roll-off is −60 dB/decade or −18 dB/octave.

Because Op Amps are seldom used in open loop for amplification of signals, we must consider the effects of feedback on the Op Amp's frequency response. Figure 6-4 shows that if the Op Amp is wired to have a closed-loop gain A_v = 10,000 or 80 dB, the bandwidth is 600 kHz. We determine this by projecting to the right of 80 dB to the point where this projection intersects the open-loop curve. Directly below this point of intersection, we read 600 kHz on the horizontal scale. Note that the open-loop curve is descending at −20 dB/decade at the point where the 80-dB projection intersects it. We can, therefore, say that the curve and projection intersect at a 20 dB/decade rate of closure.

If the Op Amp is wired for a closed-loop gain A_v = 1000 or 60 dB, we project to the right of 60 dB toward the open-loop curve. In this case, the projection and curve intersect at a 40 dB/decade rate of closure and at a point above 3.5 MHz approximately. The bandwidth *appears* to be about 3.5 MHz. However, because the open-loop curve is descending at −40 dB/decade, the circuit will likely be *unstable* and should not be used without modifications. Oscillation is a symptom of an unstable amplifier and is discussed further in later chapters. With Op Amps specifically, the main causes of instability are:

(1) In the −40 dB/decade roll-off region, the Op Amp's output signal phase shift is so large that it becomes, or approaches, an in-phase condition with the signal at the inverting input.

(2) With lower closed-loop gains, the feedback resistor R_F is relatively small compared to resistance R_1 (see Figs. 3-5 and 3-10). This increases the amount of signal that is fed back to the inverting input.

The effect of the output signal becoming more in-phase with the signal normally at the inverting input, combined with the greater amount of this output being fed back, makes the Op Amp oscillate at higher frequencies when wired to have relatively low closed-loop gain. In other words, at higher frequencies and lower closed-loop gains the feedback becomes significant and regenerative and thus meets the requirements of an oscillator.

Generally, the rate of closure between the closed-loop gain projection and the open-loop curve should not significantly exceed 20 dB/decade or 6 dB/octave for stable operation. Therefore the Op Amp with the characteristics in Fig. 6-4 is likely to be unstable if wired for gains below about 70 dB. Thus as mentioned before, if this Op Amp is wired for 60-dB gain, the rate of closure is 40 dB/decade, and the circuit will probably be unstable. Also as shown in Fig. 6-4, a gain of 20 dB causes a 60 dB/decade rate of closure which also means unstable operation is likely. When unstable, the circuit can have unpredictable output signals even if no input signal is applied.

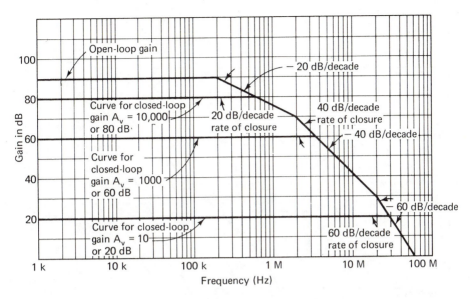

Figure 6-4

6.3 EXTERNAL FREQUENCY COMPENSATION

Some types of Op Amps are made to be used with externally connected compensating components—capacitors and sometimes resistors—if they are to be wired for relatively low closed-loop gains. These are called *uncompensated* or *tailored-response* Op Amps because the circuit designer must provide the compensation if it is required. The compensating components alter the open-loop gain characteristics so that the roll-off is about 20 dB/decade over a wide range of frequencies. For example, Fig. 6-5 shows some gain vs. frequency curves that are obtained with various compensating components on a tailored-response Op Amp. In this case, if $C_1 = 500$ pF. $C_2 = 2000$ pF, and $C_3 = 1000$ pF, then the gain vs. frequency curve 1 applies according to the table, and the roll-off starts at about 1 kHz with a steady -20 dB/decade rate. Thus if feedback components are now added to obtain an 80-dB closed-loop gain, the circuit's frequency response is flat from dc to about 3 kHz. This can be seen if we project to the right of 80 dB to curve 1. The intersection of this projection and curve 1 is at 3 kHz. Better bandwidth is obtained with curve 1 if a low closed-loop gain is used, that is, if $A_v = 100$ or 40 dB, the bandwidth increases to about 300 kHz. In either case, the rate of closure is 20 dB/decade, and the circuit is stable. Generally, if low gain is required, compensating components for curve 1 should be used. If high gain and relatively wide bandwidth are required, the compensating component for curve 3 should be used. Note in the table that if curve 3 is used, only one

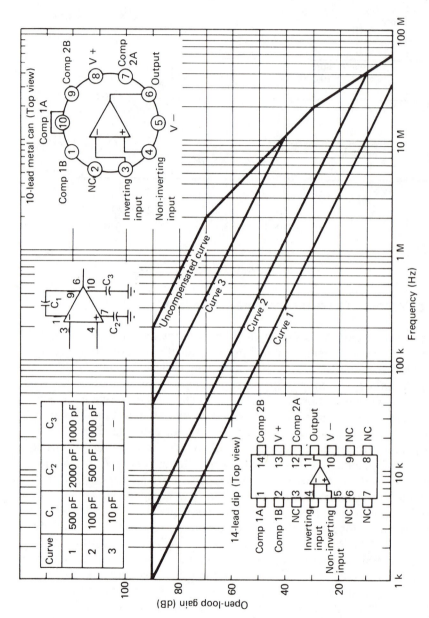

Figure 6-5 Open-loop gain vs. frequency characteristics with various externally connected (outboard) compensating components.

compensating component, $C_1 = 10$ pF, is required in this case. No compensating components need be used if this Op Amp is wired for more than 70 dB because the rate of closure never exceeds 20 dB/decade with such high gains. Note also on curve 3 that a closed-loop gain of less than 30 dB causes this gain vs. frequency response curve (projection) to meet the open-loop gain curve at a point where the rate of closure is greater than 20 dB/decade; therefore, a circuit with this gain is likely to be unstable.

Example 6-1

If the data in Fig. 6-5 applies to the Op Amp in Fig. 6-6, what are this circuit's gain V_o/V_s and bandwidth with each of the following switch positions: I, II, and III? Is this circuit stable in each switch position?

Answer.   This is a noninverting circuit and therefore its gain is

$$A_v = V_o/V_s \cong \frac{R_F}{R_1} + 1 \tag{3-6}$$

The compensating capacitors give us the gain vs frequency curve 2.
    With the switch in position I,

$$A_v \cong \frac{10 \text{ k}\Omega}{1 \text{ k}\Omega} + 1 = 11$$

This is equivalent to about 20.8 dB or about 20 dB. Projecting to the right from 20 dB in Fig. 6-5, we intersect curve 2 at 13 MHz, which means that the bandwidth in this case is about 13 MHz.
    With the switch in position II,

$$A_v \cong \frac{100 \text{ k}\Omega}{1 \text{ k}\Omega} + 1 = 101$$

which is equivalent to about 40 dB. Projecting to the right of 40 dB we intersect curve 2 at about 1.3 MHz, and therefore the bandwidth now is about 1.3 MHz.

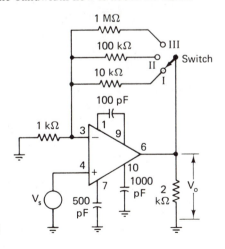

Figure 6-6

With the switch in position III,

$$A_v \cong \frac{1 \text{ M}\Omega}{1 \text{ k}\Omega} + 1 = 1001$$

which is equivalent to about 60 dB. A projection to the right of 60 dB intersects the curve 2 above 130 kHz, which means that the bandwidth now is 130 kHz.

Since all three projections intersect curve 2 at a -20 dB/decade rate of closure, the circuit is stable regardless of the switch position.

6.4 COMPENSATED OPERATIONAL AMPLIFIERS

Sometimes the relatively broad bandwidth of the uncompensated Op Amps is not needed. For example, in the instrumentation circuit shown in Fig. 5-5 of the previous chapter, the Op Amp is required to amplify relatively slow changing signals, and therefore it doesn't require good high-frequency response. In this and similar applications, internally compensated Op Amps can be used. They are sometimes simply called *compensated* Op Amps. Also, they are stable regardless of the closed-loop gain and without externally connected compensating components.

The type 741 Op Amp is compensated and has an open-loop gain vs. frequency response as shown in Fig. 6-7. The IC of the 741 contains a 30-pF capacitance that internally shunts off signal current and thus reduces the available output signal at higher frequencies. This internal capacitance,

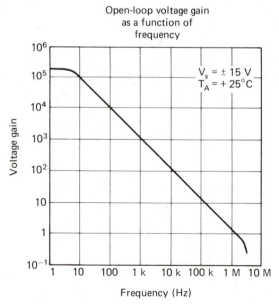

Figure 6-7 Frequency response of the 741 Op Amp.

which is an internal compensating component, causes the open-loop gain to roll off at a steady 20 dB/decade rate. Therefore, regardless of the closed-loop gain we use, the gain projection will intersect the open-loop gain curve at a 20 dB/decade rate of closure, and this assures us of a stable circuit.

The 741, along with other types of compensated Op Amps, has a 1-MHz *gain-bandwidth product*. This means that the product of the coordinates, gain and frequency, of any point on the open-loop gain vs. frequency curve is about 1 MHz.

Obviously, if the 741 Op Amp is wired for a closed-loop gain of 10^4 or 80 dB, its bandwidth is 100 Hz as can be seen by projecting to the right from 10^4 to the curve in Fig. 6-7. If the closed-loop gain is lowered, say to 10^2 or to 1, the bandwidth increases to 10 kHz or 1 MHz, respectively. The fact that the 741 has a 1-MHz bandwidth with unity gain explains why the 741 is listed with a *unity gain-bandwidth* of 1 MHz on typical specification sheets.

Example 6-2

If the Op Amp in the circuit of Fig. 6-8 is a type 741 and the circuit is required to amplify signals up to about 10 kHz, what maximum value of POT resistance can we use and still keep the circuit's response flat to 10 kHz?

Answer.   The frequency response curve in Fig. 6-7 shows that the bandwidth is greater than 10 kHz if the closed-loop gain is kept under 100. Therefore, we want to keep the ratio R_F/R_1 under 100. Thus since

$$\frac{R_F}{R_1} < 100 \text{ then } R_F < 100R_1 = 100(1 \text{ k}\Omega) = 100 \text{ k}\Omega$$

If we must keep the total feedback resistance R_F under 100 kΩ, then the POT's maximum resistance is 90 kΩ. Note the 10-kΩ fixed resistor in series with it.

6.5 SLEW RATE

An Op Amp's slew rate is related to its frequency response. Generally, we can expect Op Amps with wider bandwidths to have higher (better) slew rates. The slew rate, as mentioned in Chapter 2, is the rate of output voltage change caused by a step input voltage and is usually measured in volts per

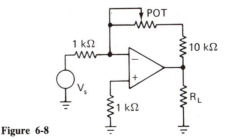

Figure 6-8

microsecond. An ideal slew rate is infinite, which means that the Op Amp's output voltage should change instantly in response to an input step voltage. Practical IC Op Amps have specified slew rates from 0.1 V/μs to about 100 V/μs which are measured in special circumstances. Some hybrid* Op Amps have slew rates on the order of 1000 V/μs. Unless otherwise specified, the slew rate listed in a data sheet was probably measured with unity gain and open load.

A less than ideal slew rate causes distortion especially noticeable with nonsinusoidal waveforms at higher frequencies. For example, ignoring overshoot, Fig. 6-9 shows typical output waveforms of a voltage follower with square waves of various frequencies applied. In this case, the slew rate is about 1 V/μs. When the input frequency is 100 Hz with the waveform in Fig. 6-9b, the output has the waveform shown in part c of the figure. Similarly, when the input is 10 kHz or 1 MHz with the waveform shown in b, the output has the waveform in Fig. 6-9d or e, respectively. Obviously, due to the limited slew rate, the squarewave input is distorted into a sawtooth at higher frequencies.

6.6 OUTPUT SWING CAPABILITY VS. FREQUENCY

An important consideration is an Op Amp's peak-to-peak output signal capability at higher frequencies. Generally, this capability decreases with higher frequencies, as shown in Fig. 6-10. In Fig. 6-10a, the type 777 Op Amp characteristics are shown for two different compensating capacitors C_c. Note that if C_c = 30 pF is used, the peak-to-peak output capability decreases sharply at operating frequencies above 10 kHz. On the other hand, when C_c = 3 pF is wired externally (outboard), the bandwidth increases, as it did with smaller compensating capacitors shown in Fig. 6-5. Also, the peak-to-peak output capability does not drop until frequencies on the order of 100 kHz or more are applied.

The output vs. frequency curve in Fig. 6-10b is for the type 741 Op Amp, which is internally compensated. Apparently then, internal compensation not only limits bandwidth, but it also limits the peak-to-peak output capability. We must keep in mind, though, that internal compensation has outstanding advantages for instrumentation and low-frequency work. For example, it simplifies our work by not forcing our attention to externally wired compensation and by assuring us of stable operation. As would be expected, faster Op Amps, intended for broadband applications, have good peak-to-peak output capability into higher frequencies (see Fig. 6-10c.)

---

* Hybrids, as opposed to monolithics (made on one chip), might contain one or more ICs or ICs and discrete components.

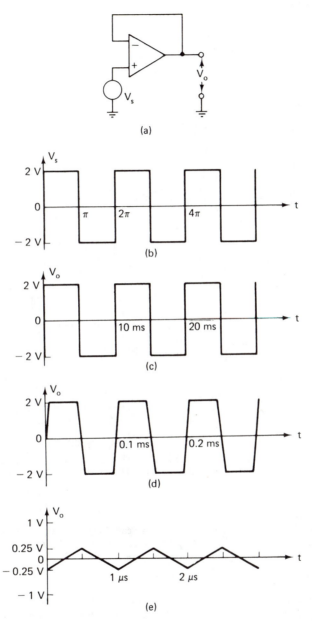

Figure 6-9 Voltage follower with an Op Amp having a slew rate of $1V/\mu s$ and squarewave voltage applied. (c) Output when the frequency of V_s is 100 Hz. (d) Output when the frequency of V_s is 10 kHz. (e) Output when the frequency of V_s is 1 MHz.

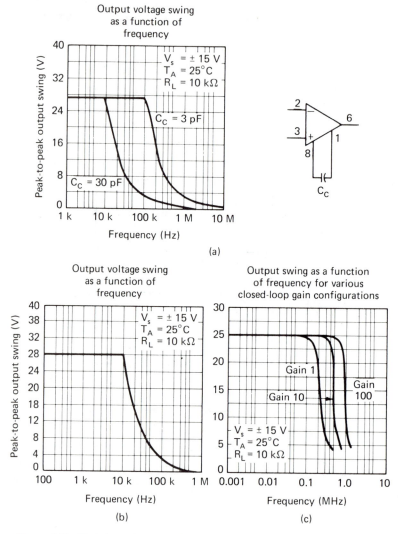

Figure 6-10 Peak-to-peak output voltage capability vs. frequency of the (a) uncompensated type 777 Op Amp; (b) compensated type 741 Op Amp; and (c) uncompensated high-speed type 715 Op Amp.

Example 6-3

If the Op Amp in the circuit of Fig. 6-8 is a type 741 with dc source voltages of ± 15 V, $R_L = 10$ kΩ, and the POT adjusted to zero, with which of the following audio-frequency input signals will this circuit clip the output signals? (Assume that the Op Amp was initially nulled and that the input signals have no dc component.)

(a) 1 Hz to 10 kHz with peaks up to ± 1 V;

(b) 1 Hz to 100 kHz with peaks up to ± 1 V;

(c) 1 Hz to 100 kHz with peaks up to ± 10 mV.

Answer. With the POT resistance 0 Ω, the total feedback resistance is 10 kΩ. Since this is an inverting amplifier, its gain $A_v = -R_F/R_1 = -10$ kΩ/1 k$\Omega = -10$. With this gain, the 741's bandwidth is about 100 kHz. See Fig. 6-7. All three signals are within the bandwidth capability of this circuit. Therefore, we can focus our attention to the peak-to-peak output capability vs. frequency.

   (a) With the 1-Hz to 10-kHz input signal peaking to ± 1 V, the output will peak to ± 10 V. Since the 741 can handle from 28 V peak-to-peak (14 V peak) up to 10-kHz, no clipping occurs.

   (b) With the 1-Hz to 100-kHz input signal peaking up to ± 1 V, the output *attempts* to peak at ± 10 V over this frequency range. However, according to the curve in Fig. 6-10b, this output is clipped with signal frequencies above 15 kHz approximately. Projecting to the right of 20 V peak-to-peak (10-V peak), we intersect the curve at roughly 15 kHz. Thus input signals with peaks of 1 V are clipped more or less, depending on how far above 15 kHz the signal is.

   (c) With the 1-Hz to 100-kHz input signal peaking up to ± 10 mV, the output peaks at ± 100 mV. No clipping occurs because the 741's peak-to-peak output capability is about 2 V or 1 V peak up to 100 kHz. This capability is well over the ± 100-mV peaks we intend to get.

6.7 HARMONICS IN NONSINUSOIDAL WAVEFORMS

The signals processed by Op Amps often are nonsinusoidal waveforms. If such waveforms are repetitive, they can be shown to consist of some *fundamental* sinewave frequency and one or more of its *harmonics*. Harmonics are also sinewaves with frequencies that are *integer* multiples of the fundamental. For example, if a nonsinusoidal waveform repeats every 1 ms, its fundamental frequency $f = 1/T = 1/1$ ms $= 1$ kHz. Its second harmonic is 2 kHz, its third harmonic is 3 kHz, etc. Examples of nonsinusoidal waveforms and their harmonic contents are shown in Fig. 6-11.

   The waveforms (b)–(d) of Fig. 6-11 are of particular interest to us. Note in these cases that the *resulting* nonsinusoidal waveforms V_R are somewhat square and that they consist of *odd*-numbered harmonics only. Especially note that the greater the number of odd harmonics, the squarer the resulting voltage V_R is. Generally, squarer waveforms contain *more* odd-numbered harmonics.

   Referring back to Fig. 6-9, we observed that the applied voltage waveform (b) is squarer than the output waveform (d). This output signal (d), therefore, contains fewer higher frequency harmonics that does the input signal (b). This is caused by the fact that an Op Amp has a limited bandwidth and cannot pass high frequencies. Thus, the higher frequency harmonics are removed or attenuated resulting in the less square output.

   Amplifiers can either remove or add harmonics from or to the signals being amplified. Fig. 6-12 shows input signals and resulting outputs of three Op Amp circuits. In circuit (a), the output V_o is a phase inverted version of

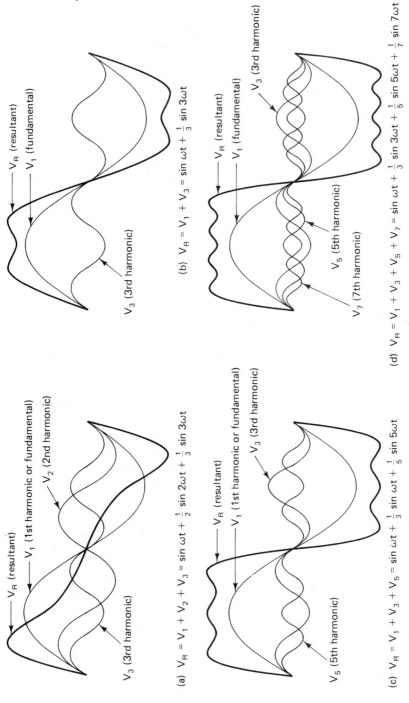

(a) $V_R = V_1 + V_2 + V_3 = \sin \omega t + \frac{1}{2} \sin 2\omega t + \frac{1}{3} \sin 3\omega t$

(b) $V_R = V_1 + V_3 = \sin \omega t + \frac{1}{3} \sin 3\omega t$

(c) $V_R = V_1 + V_3 + V_5 = \sin \omega t + \frac{1}{3} \sin 3\omega t + \frac{1}{5} \sin 5\omega t$

(d) $V_R = V_1 + V_3 + V_5 + V_7 = \sin \omega t + \frac{1}{3} \sin 3\omega t + \frac{1}{5} \sin 5\omega t + \frac{1}{7} \sin 7\omega t$

Figure 6-11 Examples of nonsinusoidal waveforms V_R and their harmonics.

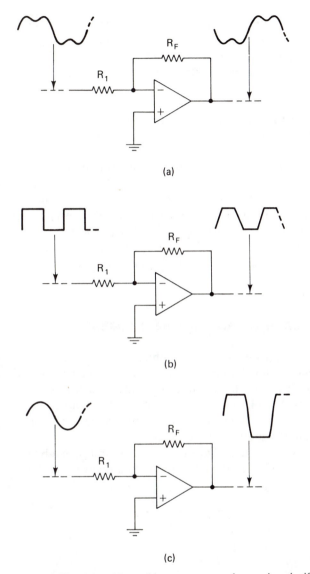

Figure 6-12 Amplifier (a) neither adds nor removes harmonics significantly; amplifier (b) removes higher frequency harmonics; and amplifier (c) adds harmonics.

the input V_s but otherwise is an undistorted signal. No significant harmonics have been added or removed. The circuit (b) has a squarewave input V_s but a less square output V_o. This output signal takes more time to increase (rise) and to decrease (fall), indicating that the signal has been stripped of its higher frequency harmonics. This amplifier, therefore, has effectively removed higher frequency components from the signal being amplified. The circuit (c) clips (distorts) the signal. Since the output is squarer than the input, this amplifier adds harmonics. The frequencies of the created harmonics are mainly odd integer multiples of the input signal frequency.

6.8 RISE AND FALL TIMES OF NONSINUSOIDAL WAVEFORMS

The terms *rise time* T_R and *fall time* T_F are commonly used to describe the characteristics of nonsinusoidal waveforms. As shown in Fig. 6-13, the rise time T_R is defined as the time it takes a voltage to rise from 10% to 90% of its peak-to-peak amplitude V_{max}. Similarly, the fall time T_F is the time it takes the waveform to fall from 90% to 10% of V_{max}. These definitions exclude the lower and upper *knees* of the waveform and, therefore, the T_R or T_F measurements are made in the most linear (straightest) portions of the rise or fall.

6.9 BANDWIDTH VS. RISE AND FALL TIMES

In previous sections we learned graphically or by brief calculations with an Op Amp's gain-bandwidth product how to determine bandwidths of Op Amp circuits. These are reliable methods but only with relatively small amplitude output signals. With larger output signals, an Op Amp's slew rate becomes a dominating factor causing the bandwidth to become narrower than the graphical (small signal) analysis indicates.

We can determine an Op Amp circuit's large signal bandwidth BW two ways. One, by driving its input with a squarewave and then observing the

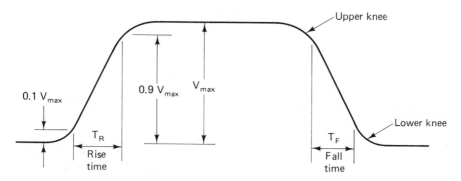

Figure 6-13 T_R is the time it takes the waveform to rise from 10% to 90% of its peak-to-peak amplitude V_{max}. T_F is the time it takes the waveform to fall from 90% to 10% of its peak-to-peak value V_{max}.

resulting output signal's rise or fall time. The *measured* T_R or T_F can then be used in the equation

$$BW \cong \frac{0.35}{T_R} = \frac{0.35}{T_F} \qquad (6\text{-}1)*$$

The second method is to calculate the rise or fall time using the Op Amp's specified slew rate. The *calculated* T_R or T_F can then be used with the above equation to determine the bandwidth BW.

Example 6-4

A squarewave, having an extremely fast rise and fall time, is applied to the input of an Op Amp circuit. Without saturating the Op Amp, the resulting output is a trapezoidal waveform, like that of Fig. 6-9d. By adjustment of the sweep time on a triggered sweep oscilloscope, we observe the leading edge of the trapezoid as shown in Fig. 6-14. If the SWEEP TIME control is in the CAL 10 μs/DIV position, what is this circuit's bandwidth?

Answer.   The rise time T_R takes about 6 horizontal divisions. Since each division is 10 μs,

$$T_R = 6(10 \ \mu s) = 60 \ \mu s$$

Using this measured T_R in Eq. (6-1) we find that the bandwidth

$$BW \cong \frac{0.35}{60 \ \mu s} \cong 5.8 \text{ kHz}$$

Example 6-5

A 741 is to output unclipped signals as large as 20 V peak to peak. What bandwidth can we expect with this output amplitude?

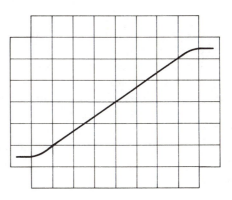

Figure 6-14   Rise time T_R measured on an oscilloscope.

Answer.   Since the 741 has a slew rate of about 0.5 V/μs, the total time T_{tot} it takes the output voltage to rise by 20 V is

$$T_{\text{tot}} = \frac{20 \text{ V}}{0.5 \text{ V}/\mu\text{s}} = 40 \ \mu\text{s}$$

The rise time T_R is about 80% of this because the first and last 10% of T_{tot} is not included in the definition of T_R. In this case, then,

$$T_R \cong 0.8(40 \ \mu\text{s}) = 32 \ \mu\text{s}$$

and, therefore, the bandwidth

$$BW \cong \frac{0.35}{32 \ \mu\text{s}} \cong 10.9 \text{ kHz} \qquad (6\text{-}1)$$

6.10 NOISE

In Chapter 5 we referred to induced 60-Hz hum as noise. Indeed, any unwanted signal mixing with desired signals is noise. Induced noise is not limited to the 60 Hz from nearby electrical power equipment. It often originates in other man-made systems such as switch arcing, motor brush sparking, and ignition systems. Natural phenomena such as lightning also cause induced noise. Induced noise voltages, as shown in Chapter 5, can be made common mode and thus can be significantly reduced relative to the desired signals at the load of an Op Amp.

The term *noise* is also commonly used to describe ac random voltages and currents generated within conductors and semiconductors. Such noise, associated with Op Amps and with amplifiers in general, limit their signal sensitivity. If very weak signals are to be amplified, very high closed-loop gain must be used to bring the signals up to useful levels. With a very high gain, however, the noise is amplified along with the signals to the point where nearly as much noise as signal appears at the output. If fed to a speaker, random noise causes a hissing, frying sound.

There are three main types of noise phenomena associated with Op Amps and with solid-state amplifiers in general. These are: *thermal* or *Johnson* noise, *shot* or *Schottky* noise, and *flicker* or 1/f noise.

Thermal noise is caused by the random motion of charge carriers within a conductor which generate noise voltages V_n within it. The rms value of this thermally generated noise voltage V_n can be predicted with the equation

$$V_n = \sqrt{4KTR(BW)} \qquad (6\text{-}2)$$

where K is Boltzmann's constant; 1.38×10^{-23} joules/°K,

T is the temperature in degrees Kelvin—the Celsius temperature plus 273°,

R is the resistance of the conductor in question, and

BW is the bandwidth in hertz.

Examining this equation, we see that thermal noise increases with higher temperatures, larger resistances, and wider bandwidths.

Although a constant average dc current may be maintained in a semiconductor, it will have random variations. These variations have an rms value referred to as noise current I_n. Noise generated in this manner is called *shot noise*. Its rms value can be predicted with the equation

$$I_n = \sqrt{2qI_{dc}(BW)} \qquad (6\text{-}3)$$

where q is the charge of an electron: 1.6×10^{-19} coulombs,

I_{dc} is the average dc current in the semiconductor, and

BW is the bandwidth.

Here again, wider bandwidths generate more noise, which is to say, we can expect less noise in narrow-bandwidth amplifiers. Of course, the noise current I_n, flowing through a resistance R, will generate noise voltage RI_n.

In addition to shot noise, semiconductors have low-frequency noise called *flicker* or $1/f$ noise. The term $1/f$ describes the inverse nature of this noise with respect to frequency, that is, the amount of flicker noise is greater at lower frequencies f.

6.11 EQUIVALENT INPUT NOISE

An Op Amp contains many active and passive components that generate and add noise to its output. These noise sources can be represented by voltage and current noise generators at the input of the amplifier's equivalent circuit, as shown in Fig. 6-15. This equivalent is sometimes called an *amplifier noise*

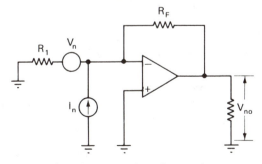

Figure 6-15 An amplifier's inherent noise voltage and current can be shown as voltage and current generators at its inputs.

model. The net effect of these input noise generators is an *equivalent input-noise voltage* V_{ni}. It is this equivalent input-noise V_{ni} that is amplified by the stage gain, and with it we can predict the output noise V_{no} that a given Op Amp stage might have.

If we neglect the thermal noise of the amplifier's signal source, the equivalent input noise V_{ni} is related to the amplifier's specified input noise voltage V_n and noise current I_n by the following equation:

$$V_{ni}^2 \cong V_n^2 + (R_E I_n)^2 \qquad (6\text{-}4)$$

where V_{ni}^2 is the mean-square equivalent input noise voltage,
V_n^2 is the specified mean-square input noise voltage,
I_n^2 is the specified mean-square input noise current, and
R_E is the parallel equivalent of R_1 and R_F.

It can be shown that equivalent input noise voltage V_{ni} sees the Op Amp circuit as a noninverting type as shown in Fig. 6-16. Broadband input noise vs. source resistance curves are sometimes shown as in Fig. 6-16b, and the general-purpose Op Amp's specified input noise voltage V_n and input noise current I_n vs. frequency characteristics are shown in Fig. 6-16c and d. Larger source resistances appreciably add to thermally generated noise as shown in part b of the figure. The $1/f$ phenomenon adds low-frequency noise as shown in Fig. 6-16c and d.

The output noise voltage V_{no} is a function of the effective input noise voltage V_{ni} and is approximated with the equation

$$V_{no}^2 \cong (A_v V_{ni})^2 \qquad (6\text{-}5a)$$

or

$$V_{no} \cong A_v \sqrt{V_n^2 + (R_F I_n)^2} \qquad (6\text{-}5b)$$

where A_v is the closed-loop gain.

REVIEW QUESTIONS

6-1. Referring to the open-loop gain of an Op Amp, what does the term *roll-off* mean?

6-2. What is a Bode diagram?

6-3. What problem can occur with an amplifier having a steep gain vs. frequency roll-off?

6-4. How can the steepness of the gain vs. *f* roll-off deliberately be limited?

6-5. Why are low closed-loop gains avoided with uncompensated Op Amps?

6-6. What advantage does the -20 dB/decade roll-off of the internally compensated Op Amp have?

6-7. What disadvantage does the internally compensated Op Amp have compared to the externally compensated types?

6-8. If an Op Amp is specified as having a *tailored response*, what does this mean?

6-9. If an Op Amp is specified as being *compensated*, what does this mean?

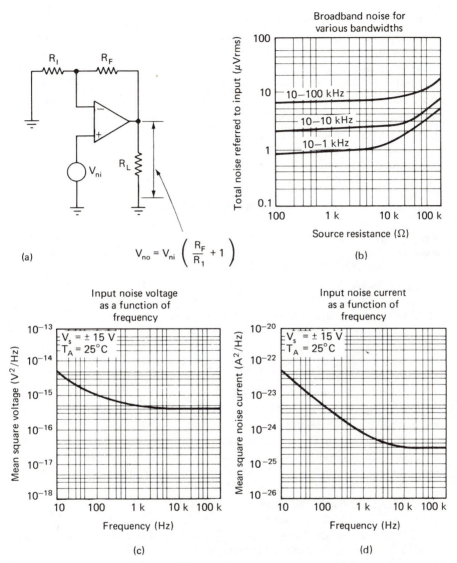

Figure 6-16 (a)

$$V_{no} = V_{ni} \left(\frac{R_F}{R_1} + 1 \right)$$

(b) Broadband noise for various bandwidths

(c) Input noise voltage as a function of frequency

(d) Input noise current as a function of frequency

Figure 6-16 The total effective input noise is amplified by the closed-loop gain of the amplifier seeing it as a noninverting type. (b) Typical noise levels higher with wider bandwidths and with larger signal source resistances. (c) The rms noise voltages squared typically developed at various frequences. (d) The rms noise currents squared typically developed at various frequencies.

6-10. If the closed-loop gain of an Op Amp is increased by increasing its feedback resistance R_F, what effect will this have on its bandwidth?

6-11. If an Op Amp's gain-bandwidth product is 2 MHz, what is its bandwidth when connected to work as a voltage follower?

6-12. What is an Op Amp's slew rate?

6-13. What effect does the operating frequency have on the maximum unclipped output signal capability of an Op Amp?

6-14. When compensating components are added externally on an Op Amp, what effect do they have on the bandwidth?

6-15. Name the three types of noise associated with conductors and semiconductors.

6-16. Generally, at what frequencies is flicker noise a greater problem?

6-17. What effect do large externally wired resistances have on the noise level at the output of an Op Amp?

PROBLEMS

Refer to Fig. 6-17a. The Op Amp will work on one of the various open-loop vs. frequency curves shown in Fig. 6-17c, depending on the values of the externally wired compensating components actually used.

6-1. For the Op Amp and characteristics of Fig. 6-17, what is the gain-bandwidth product if $C_1 = 5000$ pF, $R_1 = 1.5$ kΩ, and $C_2 = 200$ pF?

6-2. For the Op Amp and characteristics of Fig. 6-17, what is the gain-bandwidth product in the frequency range from 1 kHz to 1 MHz if $C_1 = 500$ pF, $R_1 = 1.5$ kΩ, and $C_2 = 20$ pF?

6-3. What values of compensating components should we use with an Op Amp that has characteristics of Fig. 6-17 if we need a gain-bandwidth product of 10 MHz in the range of 40-dB to 60-dB closed-loop gain?

6-4. If an Op Amp has a 10-MHz gain-bandwidth product, what is its bandwidth when used with a closed-loop gain of (a) 40 dB, and when it is (b) 60 dB?

6-5. If the Op Amp in Fig. 6-18 has the characteristics given in Fig. 6-17, what is the circuit's bandwidth when the 100-kΩ POT is adjusted to 0 Ω?

6-6. If the Op Amp in Fig. 6-18 has the characteristics given in Fig. 6-17, what is the circuit's bandwidth when the 100-kΩ POT is adjusted to maximum?

6-7. What is the bandwidth of the circuit in Fig. 6-19 if its 1-MΩ POT is adjusted to 900 kΩ and if its Op Amp has the characteristics given in Fig. 6-17?

6-8. What is the bandwidth of the circuit in Fig. 6-19, if its 1-MΩ POT is adjusted to 0 Ω and if the Op Amp has characteristics given in Fig. 6-17?

6-9. What are the approximate gain and bandwidth of the circuit in Fig. 6-20 if the 1-kΩ POT is adjusted to 0 Ω and the Op Amp has a gain-bandwidth product of 1 MHz?

6-10. If the 1-kΩ POT in the circuit of Fig. 6-20 is adjusted for maximum resistance and if the Op Amp has a gain-bandwidth product of 1 MHz, what are this circuit's gain and bandwidth?

6-11. If the Op Amp in the circuit of Fig. 6-18 has the characteristics in Fig. 6-17, what is the rate of closure between the closed-loop and the open-loop gains vs. frequency curves regardless of the adjustment of the 100-kΩ POT?

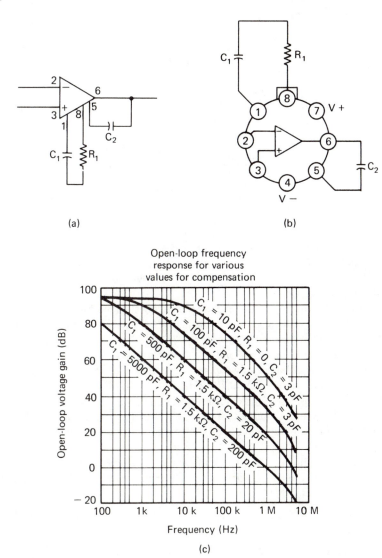

Figure 6-17 Op Amp with tailored response, and typical response curves.

6-12. In Fig. 6-20, if the Op Amp is a 741, what is the rate of closure between the closed-loop and the open-loop gains vs. frequency curves with most adjustments of the 1-kΩ POT?

6-13. The Op Amp circuit in Fig. 6-21a has the characteristics shown in Fig. 6-21b. Its input signal V_s has variable frequency and often peaks up to 80 mV. If $R_F = 100$ kΩ, beyond what applied frequency is the output signal V_o likely to be clipped? Assume that the circuit was initially nulled.

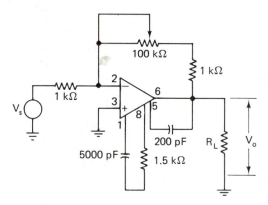

Figure 6-18

6-14. Referring to the Op Amp circuit and characteristics in Fig. 6-21, what maximum peak-to-peak voltage value of V_s can we apply and not clip the output signal V_o if $R_F = 200 \text{ k}\Omega$ and the amplifier must have a flat response up to 100 kHz?

6-15. Referring to the Op Amp and its characteristics in Fig. 6-22, how much common-mode voltage can we expect across the load R_L if the input common-mode voltage $V_{cm} = 2 \text{ mV rms}$ at 60 Hz?

6-16. Referring to the Op Amp and its characteristics in Fig. 6-22, how much common-mode voltage can we expect across the load R_L if the input common-mode voltage $V_{cm} = 2 \text{ mV}$ at 100 kHz?

6-17. Referring to Fig. 6-9, if the Op Amp circuit in part a has a slew rate of 0.5 V/μs and has the input signal V_s waveform in b of the figure applied, sketch the output waveform and indicate its peak-to-peak value if the frequency of V_s is 62.5 kHz.

6-18. If a voltage-follower circuit has the squarewave input and sawtooth output waveforms of Fig. 6-23, what is the Op Amp's slew rate? The squarewave frequency $f = 1$ MHz.

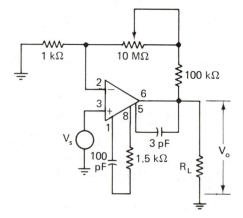

Figure 6-19

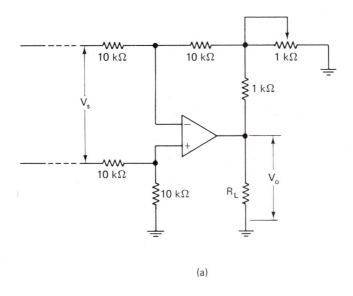

(a)

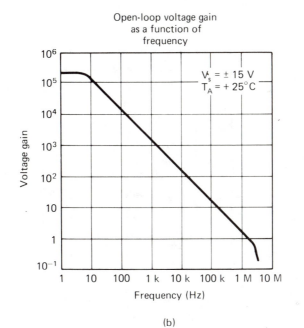

(b)

Figure 6-20

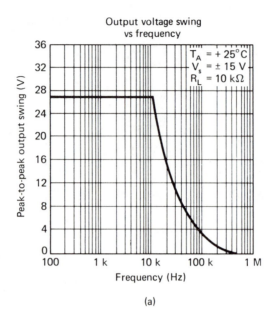

(a)

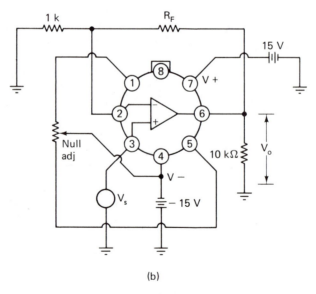

(b)

Figure 6-21

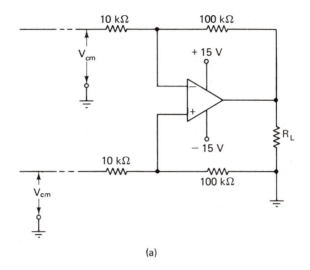

(a)

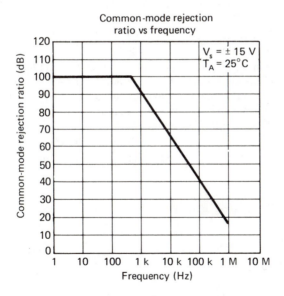

(b)

Figure 6-22

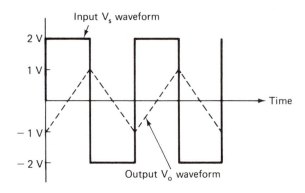

Figure 6-23

6-19. A squarewave, having an extremely fast rise and fall time, is applied to the input of an Op Amp circuit. Without saturating the Op Amp, the resulting output is a trapezoidal waveform, like (d) of Fig. 6-9. By adjustment of the sweep time on a triggered sweep oscilloscope, we observe the leading edge of the output waveform to be as shown in Fig. 6-14. If the SWEEP TIME time control is in the CAL 0.5 μs/DIV position, what is this circuit's bandwidth?

6-20. An LF351 Op Amp is to work as a voltage follower with an unclipped output signal as large as 24 V peak to peak. What bandwidth can we expect with this output amplitude? (See Appendix D.)

6-21. If the total effective input noise voltage is 5 μV rms in the circuit of Fig. 6-22, what is this circuit's output noise voltage?

6-22. If the Op Amp in the circuit of Fig. 6-18 has a specified input noise voltage of 16 nV/$\sqrt{\text{Hz}}$ and a specified input noise current of 0.01 pA/$\sqrt{\text{Hz}}$ at 1 kHz, find (a) the equivalent input noise voltage and (b) the equivalent output noise voltage at this frequency. Assume that the internal resistance of the source V_s is negligible and that the POT is set at 100 kΩ.

6-23. The POT in the circuit of Fig. 6-8 is adjusted to 90 kΩ. What is the circuit's bandwidth if the Op Amp is an LF13741?

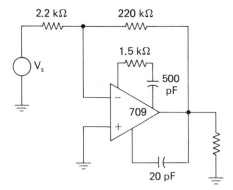

Figure 6-24

6-24. The LF351 Op Amp has a gain-bandwidth product of 4 MHz and it is pin compatible with the 741. Referring to the circuit described in the previous problem, what is its small signal bandwidth if the LF13741 is replaced with an LF351?

6-25. A voltage follower circuit has the square-wave input shown in Fig. 6-23 at 62.5 kHz. Sketch its output waveform if the Op Amp is an LF351 that has a slew rate of 13 V/μs.

6-26. The circuit of Fig. 6-24 uses a 709 Op Amp. What is its bandwidth and its gain-bandwidth product? (See the 709's Spec in Appendix H.)

6-27. Write a program in BASIC that plots the waveform V_R of Fig. 6-11a.

6.28. Write a program in BASIC that plots the waveform V_R of Fig. 6-11c.

PRACTICAL CONSIDERATIONS

In the preceding chapters we learned what an Op Amp's characteristics should be ideally and what they are in practice. In this chapter, we will see that many of the practical Op Amp characteristics, as listed on its specification sheets, are not constant and tend to drift with drifting ambient conditions. The effect of such drifting characteristics is usually most noticeable at the output of the Op Amp, and we will see methods of predicting output voltage changes vs. temperature and dc voltage source changes. In addition, we will see the characteristics of some special-purpose Op Amps and protecting techniques that are sometimes necessary in Op Amp circuits.

7.1 OFFSET VOLTAGE VS. POWER SUPPLY VOLTAGE

Because the Op Amp is capable of amplifying dc voltages, it is inherently sensitive to changes in its own dc supply voltages: the $+V$ and $-V$ sources. With practical Op Amps, if the dc supply voltages change due to poor regulation, the dc offset voltages will change too. Similarly, if the supply voltages are poorly filtered and vary at some ripple frequency, the offset voltages in the Op Amp will vary at the same frequency.

The sensitivity of an Op Amp to variations in the supply voltages is usually specified in a variety of somewhat equivalent terms such as the *power supply rejection ratio*, the *supply voltage rejection ratio*, the *power supply sensitivity*, and the *supply voltage sensitivity* to name a few. These are specified in decibels or microvolts per volt. For example,

$$\text{Power Supply Rejection Ratio, } PSRR, \text{ in decibels} = 20 \log \frac{\Delta V}{\Delta V_{io}} \qquad (7\text{-}1)$$

or

$$\text{Power Supply Sensitivity, } S, \text{ in } \mu V/V = \frac{\Delta V_{io}}{\Delta V} \qquad (7\text{-}2)$$

where ΔV is the change in the power supply voltage, and
ΔV_{io} is the resulting change in input offset voltage.

The significance of these parameters is more apparent if we recall that the input offset voltage V_{io} is the voltage required across the differential inputs to null the output of the Op Amp with no feedback (open loop). We also learned in Chapter 4 that, with feedback, the input offset voltage sees the circuit as a noninverting amplifier (see Fig. 4-5) and will cause an output offset voltage V_{oo} if the circuit has not been nulled. This output offset is therefore the stage gain times the input offset, that is,

$$V_{oo} = A_v V_{io} \qquad (4\text{-}1)$$

or

$$V_{oo} \cong \left(\frac{R_F}{R_1} + 1\right) V_{io}$$

Apparently then, any input offset voltage *change* ΔV_{io}, caused by a power supply voltage *change* ΔV, will be amplified and cause an output offset voltage *change* ΔV_{oo} that is larger by the stage gain. Equation (4-1) can therefore be modified to

$$\Delta V_{oo} \cong \left(\frac{R_F}{R_1} + 1\right) \Delta V_{io} \qquad (7\text{-}3)$$

Changes in the input offset voltage ΔV_{io} can be determined if we rearrange Eq. (7-1) or (7-2). If the power supply rejection ratio in dB is given, then

$$\Delta V_{io} = \frac{\Delta V}{\text{antilog } (PSRR(\text{dB})/20)} \qquad (7\text{-}1\text{a})$$

where the antilog of $(PSRR(\text{dB})/20)$ can quickly be estimated with the chart in Fig. 5-4. If a sensitivity factor S is given, then

$$\Delta V_{io} = S(\Delta V) \qquad (7\text{-}2\text{a})$$

Since the change in supply voltage ΔV can be due to poor regulation or poor filtering, then ΔV can be the ripple voltage V_r riding on the total supply voltage. This means that, with poor filtering, $\Delta V = V_r$ and the input offset voltage varies (changes) at the ripple frequency.

Supply voltage changes ΔV are easily measured across the $+V$ and $-V$ terminals as shown in Fig. 7-1. The meter M can be a voltmeter or an oscilloscope that is capable of measuring dc voltages and will measure ΔV caused by poor regulation or ΔV caused by poor filtering.

Example 7-1

Referring to Fig. 7-2, suppose the circuit is nulled when the voltage across terminals $+V$ and $-V$ measures 30 V dc and that, due to poor regulation, this dc voltage drifts with time from 28 V to 32 V. Also suppose that, due to poor filtering, 2.5 mV rms ac ripple is also measured across the terminals $+V$ and $-V$. While the signal source V_s = 0 V, what drift occurs in the output offset voltage and how much ripple voltage can we expect across the load R_L if (a) the Op Amp's $PSRR$ = 80 dB. Also if (b) the Op Amp's power supply sensitivity S = 150 μV/V?

Answer. Due to poor regulation, the dc supply can drift 2 V in either direction from 30 V; therefore, let ΔV = 2 V. Due to poor filtering, ΔV = 2.5 V rms.

(a) With the $PSRR$ = 80 dB, which is equivalent to a voltage ratio of 10,000 (see Fig. 5-4), then due to the drift ΔV = 2 V

$$\Delta V_{io} = \frac{2V}{\text{antilog (80 dB/20)}} = \frac{2V}{10,000} = 0.2 \text{ mV} \qquad (7\text{-}1a)$$

and the resulting output offset drift is

$$\Delta V_{oo} \cong \left(\frac{R_F}{R_1} + 1\right)\Delta V_{io} = 201(0.2 \text{ mV}) = 40.2 \text{ mV} \qquad (7\text{-}3)$$

Due to the ripple ΔV = 2.5 mV rms,

$$\Delta V_{io} = \frac{2.5 \text{ mV}}{10,000} = 0.25 \ \mu\text{V rms} \qquad (7\text{-}1a)$$

and the output ripple voltage is

$$V_{o(\text{ripple})} \cong \left(\frac{R_F}{R_1} + 1\right)\Delta V_{io} = 201(0.25 \ \mu\text{V}) = 50.25 \ \mu\text{V rms} \qquad (7\text{-}3)$$

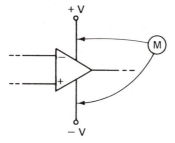

Figure 7-1 Meter M reads the total dc supply voltage.

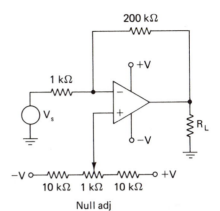

Figure 7-2

(b) With the sensitivity factor $S = 150\ \mu V/V$, which is equivalent to 150×10^{-6}, and a dc supply drift $\Delta V = 2$ V,

$$\Delta V_{io} = S(\Delta V) = 150 \times 10^{-6}(2\ V) = 300\ \mu V = 0.3\ mV \qquad (7\text{-}1b)$$

and the resulting output offset drift

$$V_{oo} \cong 201(0.3\ mV) = 60.3\ mV \qquad (7\text{-}3)$$

Due to the ripple $\Delta V = 2.5$ mV rms,

$$\Delta V_{io} = 150 \times 10^{-6}(2.5\ mV) = 0.375\ \mu V\ rms \qquad (7\text{-}1b)$$

and this causes an output ripple voltage of

$$V_{o(ripple)} \cong 201(0.375\ \mu V) \cong 75.4\ \mu V\ rms \qquad (7\text{-}3)$$

Sometimes a drift in input bias current vs. dc supply voltage is specified on manufacturers' data sheets in terms of picoamperes per volt. For example, an input bias current vs. dc supply characteristic might be specified as ± 10 pA/V, which means that the input bias current might either increase or decrease by as much as 10 pA for every one-volt change in the dc supply as measured across the $+V$ and $-V$ terminals. Any drift in the input bias current will tend to cause a drift in the output offset, which is discussed in more detail in the next section.

7.2 BIAS AND OFFSET CURRENTS VS. TEMPERATURE

The typical and maximum values of input bias current I_B and input offset current I_{io}, as specified in manufacturers' data sheets, are usually values measured at 25°C, which is about room temperature. We usually cannot depend on these specified currents holding to steady values if the temperature of the Op Amp drifts. Typical I_B vs. temperature and I_{io} vs. temperature curves for a general-purpose IC Op Amp are shown in Fig. 7-3. It is interesting to note that these bias and offset currents increase with decreased

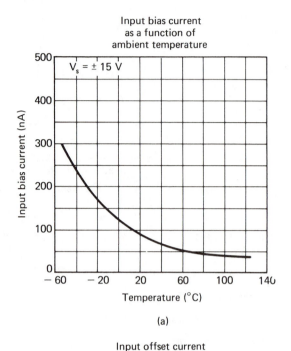

(a)

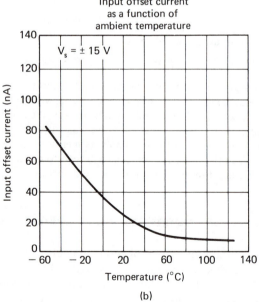

(b)

Figure 7-3 General-purpose Op Amp curves: (a) input bias current I_B vs. temperature; (b) input offset current I_{io} vs. temperature.

temperatures, which is generally the case with transistor-input IC Op Amps. The input bias current rises with higher temperatures in FET-input types of Op Amps. In either case, temperature changes cause bias and offset current changes, resulting in a drift in the output offset voltage. High-performance Op Amps are available that feature extremely low drift with temperature changes. As would be expected, their costs are higher than for the general-purpose types. The point is, we should be aware of the economical IC Op Amp's tendency to drift. If the drift is excessive for the application we have in mind, we can then consider the higher-performance types.

We concern ourselves with either the drift in input bias current ΔI_B or with the drift in the input offset current ΔI_{io}, depending on how the circuit is wired. Generally, if an Op Amp's dc resistance to ground, as seen from its noninverting input 2, is negligible compared to the dc resistance to ground, seen from the inverting input 1, we can closely estimate the circuit's change or drift in output offset voltage with the equation

$$\Delta V_{oo} \cong R_F \Delta I_B \qquad (7\text{-}4)$$

where R_F is the feedback resistance across the inverting input 1 and the output, and

ΔI_B is the change or drift in the input bias current.

If the dc resistances to ground as seen from both the inverting and noninverting inputs are equal, the change or drift in the output offset voltage can be closely estimated with the equation

$$\Delta V_{oo} \cong R_F \Delta I_{io} \qquad (7\text{-}5)$$

where ΔI_{io} is the change or drift in the input offset current.

For example, in Fig. 7-4a, the noninverting input 2 of the circuit is grounded; therefore, Eq. (7-4) is used to predict the drift in output offset ΔV_{oo}. Equation (7-4) also applies if R_2 is much smaller than $R_1 + R_s$ in the circuit of Fig. 7-4b and if R_s is much smaller than R_1 in the circuit in c of this figure.

On the other hand, if

$$R_2 = \frac{(R_1 + R_s)R_F}{R_1 + R_s + R_F} \qquad (7\text{-}6)$$

in the circuit of Fig. 7-4b, and if

$$R_s = \frac{R_1 R_F}{R_1 + R_F} \qquad (7\text{-}7)$$

in the circuit of Fig. 7-4c, the dc resistances to ground seen looking from both inputs of each circuit are equal, and Eq. (7-5) applies.

Since ΔI_{io} is much smaller than the ΔI_B of a given Op Amp, designing for equal dc resistances to ground reduces an Op Amp circuit's drift. All the circuits of Fig. 7-5 are low-drift (stable) designs, provided their component

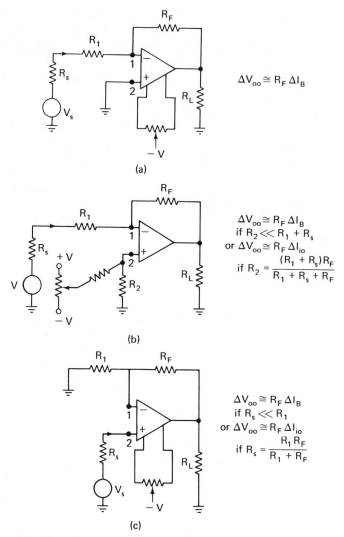

Figure 7-4   There is a change in output offset voltage ΔV_{oo} if a change in the input bias current ΔI_B occurs.

values are properly selected. If R_2 is selected with Eq. (7-6) in the circuit of Fig. 7-5a, and if $R_a = R_1$ and $R_b = R_F$ in the circuit b of the figure, and if

$$R_2 = R_s - \frac{R_1 R_F}{R_1 + R_F} \tag{7-8}$$

in the circuit of Fig. 7-5c, the output offset drift ΔV_{oo} is closely related to the offset current drift ΔI_{io} rather than bias current drift ΔI_B which make these circuits less prone to drift with temperature changes.

Example 7-2

(a) The Op Amp in the circuit of Fig. 7-4b has the characteristics of Fig. 7-3 and is nulled at 25°C. What output offset is across the load R_L at 60°C if $R_s = 0\ \Omega$, $R_1 = 10$ kΩ, $R_F = 1$ MΩ, and $R_2 = 100\ \Omega$?

(b) The Op Amp in the circuit of Fig. 7-5b has the characteristics shown in Fig. 7-3 and is nulled at 25°C. What output offset is across the load R_L at 60°C if $R_a = R_1 = 10$ kΩ and $R_b = R_F = 1$ MΩ?

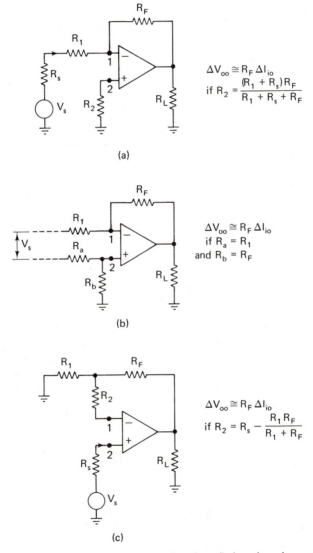

(a)

$$\Delta V_{oo} \cong R_F \Delta I_{io}$$
$$\text{if } R_2 = \frac{(R_1 + R_s)R_F}{R_1 + R_s + R_F}$$

(b)

$$\Delta V_{oo} \cong R_F \Delta I_{io}$$
$$\text{if } R_a = R_1$$
$$\text{and } R_b = R_F$$

(c)

$$\Delta V_{oo} \cong R_F \Delta I_{io}$$
$$\text{if } R_2 = R_s - \frac{R_1 R_F}{R_1 + R_F}$$

Figure 7-5 In stabilized circuits (R_2 properly selected) there is a change in the output offset voltage ΔV_{oo} if a change in the input offset current ΔI_{io} occurs.

Answer. (a) Since $R_2 \ll R_1 + R_s$, Eq. (7-4) and the curve of Fig. 7-3a apply. On this curve, $I_B \cong 80$ nA at 25°C, but it drops to 50 nA at 60°C. Thus, $\Delta I_B \cong 80 - 50 = 30$ nA. This causes a

$$\Delta V_{oo} \cong R_F \Delta I_B \cong 1 \text{ M}\Omega(30 \text{ nA}) = 30 \text{ mV}$$

which means that the output voltage V_o drifts from its initial 0 V at 25°C to 30 mV as the temperature rises to 60°C.

(b) In this case, since $R_a = R_1$ and $R_b = R_F$, Eq. (7-5) and the curve in Fig. 7-3b apply. On this curve, $I_{io} \cong 25$ nA at 25°C, and it decreases to about 10 nA at 60°C. Therefore, with the $\Delta T = 60 - 25 = 35°C$, the $\Delta I_{io} \cong 25 - 10 = 15$ nA. The resulting change in output offset voltage is

$$\Delta V_{oo} \cong R_F \Delta I_{io} \cong 1 \text{ M}\Omega(15 \text{ nA}) = 15 \text{ mV}$$

Thus as the temperature increases from 20°C to 60°C, the output offset voltage increases from its initial 0 V to 15 mV.

Sometimes the drift in input bias current due to temperature changes is specified in terms of picoamperes per degree Celsius over a specified temperature range. For example, some varactor-input Op Amps have a specified drift of 0.001 pA/°C over the temperature range from +10°C to 70°C. Some FET-input Op Amp's data sheets might show that current simply doubles for every 10° rise in temperature in the range from −25°C to +85°C.

Needless to say, specifications on drift, whether on curves or in tables, enable us to predict the maximum drift we might have in the output voltage level under known temperature variations.

7.3 INPUT OFFSET VOLTAGE VS. TEMPERATURE

The input offset voltage V_{io}, as listed on the manufacturers' data sheets, was measured at 25°C unless otherwise specified. If the temperature changes, the input offset voltage changes too. The data sheets usually specify this drift (change) in microvolts per degree Celsius (μ V/°C). We already noted, in Eq. (7-3), that a change in the input offset voltage causes a change in the output voltage of the Op Amp.

Example 7-3

In the circuit of Fig. 7-5b, $R_a = R_1 = 10$ kΩ and $R_b = R_F = 1$ MΩ. It was nulled at 25°C. If the Op Amp has a specified $\Delta V_{io}/\Delta T = 15$ μV/°C, what is the output offset at 60°C as caused by the drift in V_{io}?

Answer. In this case, the temperature change

$$\Delta T = 60°C - 25°C = 35°C$$

Therefore, the resulting change in the input offset voltage

$$\Delta V_{io} = (15 \text{ } \mu\text{V/°C})35°C = 525 \text{ } \mu \text{ V}$$

By Eq. 7-3,

$$\Delta V_{oo} = \left(\frac{R_F}{R_1} + 1\right)\Delta V_{io} = \left(\frac{1\ \text{M}\Omega}{10\ \text{k}\Omega} + 1\right)525\ \mu\text{V} \cong 53\ \text{mV}$$

7.4 OTHER TEMPERATURE-SENSITIVE PARAMETERS

A drift in the Op Amp's output voltage level is not the only possible result of temperature changes. Figure 7-6 shows a typical collection of other temperature-sensitive characteristics. The curves in Fig. 7-6a show the drift of three parameters with temperature changes on a relative scale. In this case, relative scale indicates multiplying factors that enable us to determine the changes in the three parameters if the temperature drifts away from 25°C. For example, Fig. 7-6a shows that the transient response at 70°C is 1.1 times the transient response at 25°C. On this same graph, the closed-loop *BW* at 70°C is shown to be 0.9 of its value at 25°C. The slew rate is shown to be quite constant, especially with temperatures above 25°C.

The transient response of an Op Amp is the time (usually in microseconds) required for its output voltage to rise from 10% to 90% of its final value under *small-signal* conditions. A typical transient response waveform is shown in Fig. 7-7. Note that the voltage level here is much smaller than that used to define the slew rate.

As shown in Fig. 7-6b, the input resistance R_i of the typical Op Amp increases with an increasing temperature. This could be a problem if we drive a noninverting amplifier with a high internal resistance signal source. A drifting input resistance, which is the load resistance on the signal source, can cause a drifting input signal amplitude.

7.5 CHANNEL SEPARATION

If a monolithic Op Amp package contains two amplifiers (dual Op Amp), a parameter called *channel separation* is specified in its data sheets. It tells us how much interaction we can expect between the two amplifiers. If a signal is applied to the input of one amplifier in a dual package, some signal will appear at the output of the other Op Amp even though it has no input signal applied. This interaction exists because of the close physical proximity, which causes electrical coupling, between the two Op Amps built on a single semiconductor chip.

The 747, whose packages are shown in Fig. 7-8, contains essentially two 741 Op Amps. Its channel separation is typically specified at a minimum of 100 dB. This means that if one of the Op Amps is driven while the other is not, the output signal of the undriven one will be at least 100 dB below the signal output of the driven amplifier.

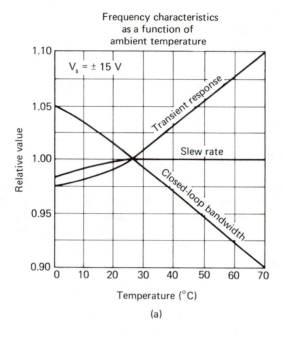

(a)

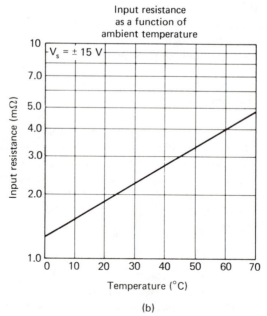

(b)

Figure 7-6 Temperature-sensitive parameters of the typical Op Amp.

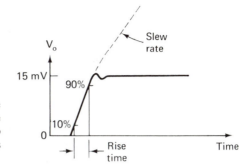

Figure 7-7 Typical transient response waveform. The rise time is the time the output voltage can change from 10% to 90% of its final level under small-signals conditions.

Example 7-3

If one of the Op Amps of a 747 is driven so that its output signal is 15-V peak to peak at 400 Hz, how much 400-Hz signal might be at the output of the other amplifier even though it is not driven?

Answer.   Since the 747's channel separation is at least 100 dB, which is equivalent to a ratio of 10^5, the output of the undriven amplifier will not be more than

$$\frac{15 \text{ V}}{10^5} = 150 \text{ } \mu\text{V peak to peak}$$

7.6 CLEANING PC BOARDS AND GUARDING INPUT TERMINALS

As mentioned previously, some applications require Op Amps with low bias currents. Certain types of high-beta transistor-input Op Amps, and FET-input Op Amps are made for such applications. Use of such low-bias-current Op Amps presents new problems, however. When the low-bias-current Op Amp is used on a printed circuit (PC) board, the circuit tends to behave erratically and may drift with temperature changes, especially in higher temperature ranges, if certain precautions are not taken.

A drift problem can be caused by PC board leakage currents. These currents can easily exceed the bias currents of the Op Amp, especially if it is a low-bias-current type. To reduce the possibility of such drift, the printed circuit board must be thoroughly cleaned with alcohol or trichlorethylene to remove all solder flux. The board can then be coated with silicone rubber or epoxy to keep its surface clean.

Even a clean PC board has finite resistance. At 125°C, the resistance between parallel conductors 1 inch long with 0.05-inch separation is about 10^{11} Ω. With just a few volts across such conductors, the leakage current between them will easily run up to hundreds of picoamperes, which is larger than bias currents in some high-performance Op Amps. Therefore, because of the close spacing between pins, the dc supply voltages can force currents

Dual-in-line packages

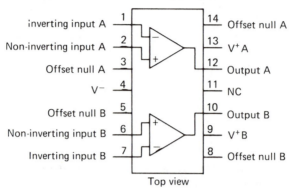

Top view

Metal can package

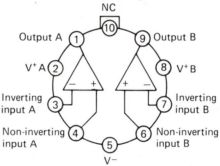

Top view

Flat package

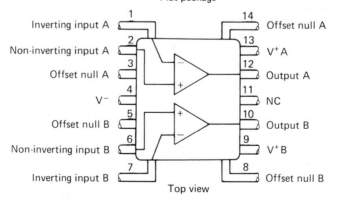

Top view

Figure 7-8   Packages containing dual Op Amps.

through the board and Op Amp's inputs. Since such currents might be large relative to the Op Amp's bias current, they can drive the output into saturation or at least cause a drifting output voltage. This problem is reduced by surrounding the inputs with a conductive *guard,* which is terminated to some low-impedance point that is essentially at the same potential as the inputs. This effectively prevents current flow through the board material into the Op Amp's inputs.

Figure 7-9 shows a PC board layout with a guard around the inputs, along with its schematic representation and proper termination in popular circuit types.

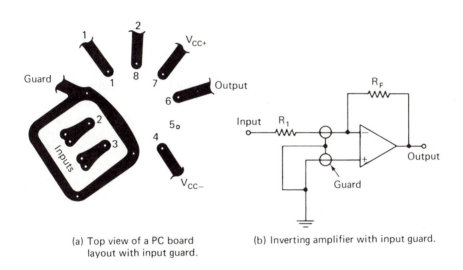

(a) Top view of a PC board layout with input guard.

(b) Inverting amplifier with input guard.

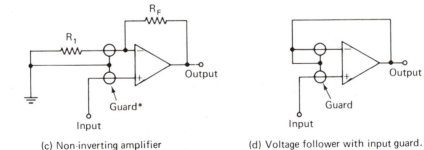

(c) Non-inverting amplifier with input guard.

(d) Voltage follower with input guard.

* This guard should be tied to a dc voltage equal to the dc average input signal voltage.

Figure 7-9   (a) Top view of a PC board layout with input guard; (b) inverting amplifier with input guard; (c) noninverting amplifier with input guard; (d) voltage follower with input guard.

7.7 PROTECTING TECHNIQUES

It takes on the order of just a few millivolts across an Op Amp's inverting and noninverting inputs to drive its output into saturation. Large differential inputs can ruin it. If large input peaks or transients are expected across the inputs, the Op Amp can be protected as shown in Fig. 7-10. The diodes remain nonconducting and therefore do not affect the input signals as long as these signals are small as they normally should be. Large input signals, however, drive the diodes into conduction. Thus the differential input voltages are limited to a few hundred millivolts—the forward voltage drop of each diode.

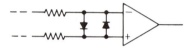

Figure 7-10 Input breakdown protection.

Most IC Op Amps have built-in *output short-circuit protection*. Note that this quality is specified in Appendix F. Some types, such as the 702 and the 709, can tolerate an output short circuit for just a short time and can deliver about 75 mA. If we keep such current drain for too long, the unprotected Op Amp will be ruined. A resistor in series with the output, as in Fig. 7-11, will keep the current drain within safe limits. For the 709, the manufacturers recommend a 200-Ω series output resistor if the output can become short-circuited to ground. Of course, if a fixed load resistor greater than 200 Ω is used, there is no need for using a series output resistor even though the Op Amp is not internally protected.

Op Amps can be ruined if the dc supply voltages are connected in the wrong polarities. Diodes in series with the dc supply leads, such as in Fig. 7-12, will protect the Op Amp from such an accident.

If unregulated or poorly regulated dc supplies are used, the Op Amp can be protected from excessive supply voltage with a zener (regulator diode) as shown in Fig. 7-13. Thus if the Op Amp's maximum dc supply voltages are specified at ± 17 V, a zener with a breakdown voltage of 34 V or less would be used.

REVIEW QUESTIONS

7-1. If a hypothetical Op Amp has an infinite power supply rejection ratio, how stringent should its power supplies' regulation and filtering be?

7-2. What are some causes of drift in the output voltage level of an Op Amp with temperature changes?

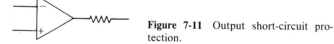

Figure 7-11 Output short-circuit protection.

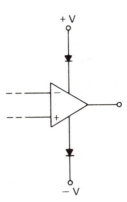

Figure 7-12 Dc supply voltage reversal
protection.

7-3. From which would we expect more drift: an Op Amp with relatively large or a relatively low input bias current? Why?

7-4. In the circuit of Fig. 7-4a, if given the change in the input bias current and the change in the input offset current with temperature, which of these specifications enables us to predict the output voltage drift? Why?

7-5. In the circuit of Fig. 7-5a, if we know the Op Amp's change in input bias current and its change in input offset current with temperature, which of these specifications enables us to predict the output voltage drift? Why?

7-6. Name five Op Amp parameters that drift with temperature changes.

7-7. Compare the parameters of Op Amp types LM741C and LF351A. Which would you expect to have more drift? Why? (See Appendix D.)

7-8. What is the purpose of resistor R_2 in the circuit (c) of Fig. 7-5?

7-9. What PC board precautions should be followed when Op Amps having extremely low bias currents are used?

7-10. If the ambient temperature is subject to considerable changes, which is more stable, with a given type of general-purpose Op Amp, the circuit in Fig. 7-4a or the circuit in Fig. 7-5a? Why?

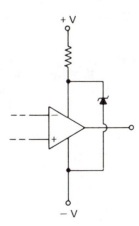

Figure 7-13 Dc supply overvoltage protection.

PROBLEMS

7-1. In the circuit of Fig. 7-2, suppose that we null the output when the voltage across the $+V$ and $-V$ supply terminals measures 24 V. This voltage then drifts from 20 V to 28 V. If the Op Amp's *PSRR* = 100 dB, what changes can we expect in its (a) input offset voltage, and (b) output voltage level?

7-2. In the circuit of Fig. 7-5b, what drift in input offset voltage and output voltage can we expect if $R_1 = R_a = 10$ kΩ, $R_F = R_b = 1$ MΩ, the power supply sensitivity $S = 100$ μV/V, and the voltage across the $+V$ and $-V$ supply terminals drifts in the range from 20 V to 25 V?

7-3. In the circuit described in Problem 7-1, how much 60-Hz hum can we expect across the load if there is a 4-mV rms, 60-Hz voltage across the $+V$ and $-V$ pins?

7-4. In the circuit described in Problem 7-2, how much 60-Hz hum can we expect across the load if a 6-mV rms, 60-Hz voltage is measured across the $+V$ and $-V$ terminals?

7-5. If an Op Amp has a *PSRR* of 100 dB, what is its power supply sensitivity S in μ V/V?

7-6. If an Op Amp has a power supply sensitivity $S = 200$ μ V/V, what is its *PSRR* in dB?

7-7. In the circuit of Fig. 7-4a, suppose that $R_s = 1000$ Ω, $R_1 = 20$ kΩ, $R_F = 10$ MΩ, and the circuit is nulled at 25°C. What dc output could we expect at 65°C if the Op Amp is an FET type whose bias current is 30 pA at 25°C and whose bias current doubles for every 10°C rise in temperature?

7-8. In the circuit of Fig. 7-4a, suppose that $R_s = 100$ Ω, $R_1 = 1$ kΩ, $R_F = 20$ kΩ, and the circuit is nulled at 20°C. What dc output voltage can we expect at 0°C if the Op Amp has the characteristics shown in Fig. 7-3?

7-9. If in the circuit of Fig. 7-5a, $R_s = 200$ Ω, $R_1 = 2$ kΩ, $R_2 = 2$ kΩ, $R_F = 20$ kΩ, and the dc output is 500 mV at 20°C, what is the approximate maximum possible output offset voltage at (a) 0°C and at (b) 60°C? Assume that the Op Amp has characteristics as shown in Fig. 7-3.

7-10. Assume that in the circuit of Fig. 7-4b, $R_1 = 1$ kΩ, $R_2 = 1$ kΩ, $R_s = 100$ Ω, and $R_F = 10$ kΩ. If the Op Amp has the characteristics shown in Fig. 7-3, and if the circuit is nulled at 20°C, what dc output can we expect at temperatures (a) -20°C and (b) 100°C?

7-11. If the FET-input Op Amp in the circuit described in Prob. 7-7 has a $\Delta V_{io}/\Delta T = 20$ μ V/°C, (a) what drift in output offset could we expect at 65°C as caused by the drift in V_{io}? (b) What is the drift in output offset as caused by the combined effects of the drifts in bias current and input offset voltage?

7-12. If in the circuit of Fig. 7-4a, $R_s = 0$, $R_1 = 2.2$ kΩ, $R_F = 820$ kΩ and the Op Amp is an LF351 type, what drift in the output voltage can we expect if the temperature can vary from 25°C to 65°C? (See Appendix D for Specs on the LF351.)

7-13. If an Op Amp has the characteristics shown in Fig. 7-6a and has a 100-ns transient response at 25°C, what is its transient response at (a) 0°C and at (b) 70°C?

7-14. If an Op Amp has the characteristics shown in Fig. 7-6a and has a closed-loop bandwidth of 40,000 Hz at 25°C, what are its bandwidths at temperatures (a) 0°C and (b) 70°C?

7-15. If $R_1 = R_a = 1.1$ kΩ and $R_F = R_b = 680$ kΩ in the circuit of Fig. 7-5b, what is the output offset voltage V_{oo} as caused by bias current (a) if the Op Amp is a 741 type, and (b) if the Op Amp is an LF351 type? Note: The LF351 is a FET-input Op Amp that is pin compatable with the 741. (See Appendix D.)

7-16. Find the value of R_2 in the circuit of Fig. 7-5c if $R_1 = 1$ kΩ, $R_F = 10$ kΩ and the internal resistance of the signal source $R_s = 40$ kΩ. With your value of R_2 in this circuit, what is its closed-loop voltage gain?

8

ANALOG APPLICATIONS
OF OP AMPS

Electronic circuits are classified as being either analog or digital. Op Amps are made to work as analog devices, though as we will see later they can perform in digital circuits too. An analog circuit has a continuously variable output voltage or current that is a function of (mathematically related to) an input voltage or current. All amplifiers described in the previous chapters are analog applications of Op Amps. Amplifiers, more specifically, fall into a category called *linear circuits* which is a subset of analog circuits. Generally, a linear circuit has an output voltage or current that is proportional to (a multiple of) an input voltage or current. Needless to say, an Op Amp can work in an analog mode only if its output voltage and frequency are within its capabilities. An overview of previously described circuits is given here along with discussions of other common analog applications of Op Amps.

8.1 OP AMPS AS AC AMPLIFIERS

The Op Amp applications discussed till now were mainly dc amplifiers. In other words, their output voltages change in respose to changes in dc input levels. All the circuits in Fig. 8-1 were discussed previously and are dc type amplifiers. Of course, they respond to ac input signals too, provided the frequencies are not too high. In some applications, the circuit designer needs the ac response characteristics of the Op Amp but does not need or want its dc response capability. A multistage audio amplifying system is an example. Even a small dc input offset voltage in the first stage can easily be amplified

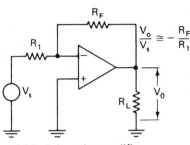

$$\frac{V_o}{V_s} \cong -\frac{R_F}{R_1}$$

(a) Basic inverting amplifier.

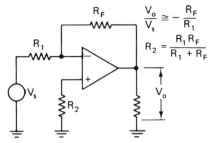

$$\frac{V_o}{V_s} \cong -\frac{R_F}{R_1}$$

$$R_2 = \frac{R_1 R_F}{R_1 + R_F}$$

(b) Inverting amplifier with improved stability.

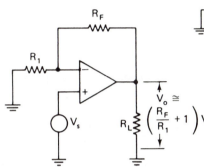

$$V_o \cong \left(\frac{R_F}{R_1} + 1\right) V_s$$

(c) Basic non-inverting amplifier.

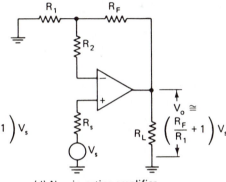

$$V_o \cong \left(\frac{R_F}{R_1} + 1\right) V_s$$

(d) Non-inverting amplifier with improved stability.

$$R_2 = R_s - \frac{R_1 R_F}{R_1 + R_F} \quad \text{if} \quad R_s > \frac{R_1 R_F}{R_1 + R_F}$$

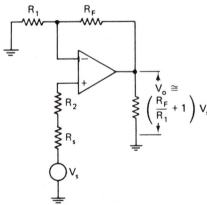

$$V_o \cong \left(\frac{R_F}{R_1} + 1\right) V_s$$

(e) Non-inverting amplifier with improved stability.

$$R_2 = \frac{R_1 R_F}{R_1 + R_F} - R_s$$

$$\text{if} \quad R_s < \frac{R_1 R_F}{R_1 + R_F}$$

$$V_o \cong \frac{R_F}{R_1} V_s$$

(f) Basic differential input amplifier.

$R_1 = R_a$ and $R_F = R_b$
for good CMRR

Figure 8-1   Common linear Op Amp applications.

to a value large enough to saturate the following Op Amp stages if the stages are directly coupled. Capacitive coupling, as shown in Fig. 8-2, between stages is a simple way of eliminating dc level amplification from stage to stage. The signal source V_s could be a preceding amplifier with a dc component. The coupling capacitor C effectively blocks this dc component, thus preventing it from affecting the next stage. The resistor R_2 *must* be used because it provides a dc path between the noninverting input and ground. Without R_2, the coupling capacitor C would become charged by the bias current I_{B_2} and would cause a dc input voltage on the noninverting input, resulting in a dc output offset and usually saturation.

Since the circuit in Fig. 8-2 is basically a noninverting amplifier, its closed-loop gain is

$$A_v \cong \frac{R_F}{R_1} + 1 \qquad (3\text{-}6)$$

at frequencies within its bandwidth. The low-frequency limit of the bandwidth is determined largely by the input resistance and the size of the coupling capacitor C.*

An inverting ac amplifier is shown in Fig. 8-3. If the reactance of C is negligible, its gain is

$$A_v \cong -\frac{R_F}{R_1} \qquad (3\text{-}4)$$

at frequencies within its bandwidth. If high gain or a large feedback resistor R_F is used, a resistor R_2 equal to R_F is used between the noninverting input and ground to reduce the output offset and drift caused by bias current.

With negligible reactances of the coupling capacitors C, the signal source V_s sees R_2 as the stage input impedance in the circuit of Fig. 8-2, and

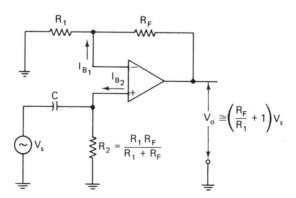

Figure 8-2   An ac noninverting amplifier; I_{B_1} and I_{B_2} are dc bias currents.

* See Appendix L for a method of selecting the proper coupling capacitor values.

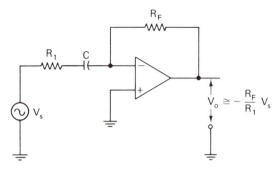

Figure 8-3   Ac inverting amplifier.

it sees R_1 as the input resistance in the circuit of Fig. 8-3. These impedances are relatively low and in cases where the ac signal source V_s must see a very large stage input impedance, the amplifier's input impedance can be *bootstrapped* as shown in Fig. 8-4. In this circuit, which is a voltage follower, the output signal V_o is applied to the bottom of R_1 via capacitor C_2. Simultaneously the input signal V_s is applied to the top of R_1 through the coupling capacitor C_1. Therefore, the signal potential difference across R_1 is $V_s - V_o$. But since $V_s \cong V_o$ in this voltage follower, their difference is extremely small, which means that the signal current through R_1 is also extremely small. Since the signal current in R_1 is also the current drain from the signal source V_s, the impedance seen by the signal source V_s is extremely large.

8.2 SUMMING AND AVERAGING CIRCUITS

The circuit in Fig. 8-5 is the basic inverting summing and averaging circuit. The value and polarity of the output voltage V_o is determined by the sum of the input voltages V_1 through V_n and by the values of the externally wired resistors. Since the noninverting input is at ground potential, the inverting

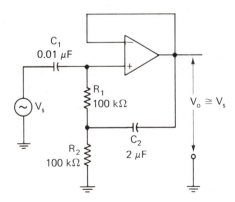

Figure 8-4 Ac voltage follower with in-put impedance bootstrapped (increased).

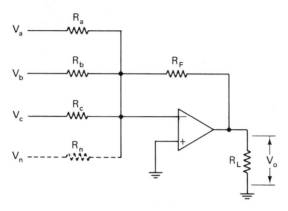

Figure 8-5　Basic inverting summing or averaging circuit.

input and the right sides of input resistors R_a through R_n are virtually grounded too. Thus, the currents in the input resistors can be shown, by Ohm's law, as

$$I_a \cong \frac{V_a}{R_a}$$

$$I_b \cong \frac{V_b}{R_b}$$

$$I_c \cong \frac{V_c}{R_c}$$

$$I_n \cong \frac{V_n}{R_n}$$

Similarly, the current in R_F is

$$I_F \cong \frac{-V_o}{R_F}$$

where the negative sign indicates that V_o is out of phase with the net voltage at the inverting input. Since the sum of the input currents is equal to the feedback current, assuming that the Op Amp is ideal, we can show that

$$I_F \cong I_a + I_b + I_c + \cdots + I_n$$

or

$$\frac{-V_o}{R_F} \cong \frac{V_a}{R_a} + \frac{V_b}{R_b} + \frac{V_c}{R_c} + \cdots + \frac{V_n}{R_n}$$

Multiplying both sides of the above equations by R_F shows that, generally, the output voltage is

$$V_o \cong -R_F \left(\frac{V_a}{R_a} + \frac{V_b}{R_b} + \frac{V_c}{R_c} + \cdots + \frac{V_n}{R_n} \right) \qquad (8\text{-}1)$$

If all input resistors are equal, say

$$R_a = R_b = R_c = R_n = R$$

Eq. (8-1) simplifies to

$$V_o \cong -\frac{R_F}{R} (V_a + V_b + V_c + \cdots + V_n) \qquad (8\text{-}2)$$

If $R_F = R$, this equation simplifies further to

$$V_o \cong -(V_a + V_b + V_c + \cdots + V_n) \qquad (8\text{-}3)$$

This last equation shows that the circuit in Fig. 8-5 can be used to find the negative sum of any number of input voltages.

This circuit can also average the input voltages by using a ratio of R_F/R that is equal to the reciprocal of the number of voltages being averaged. The ratio R_F/R is selected so that the sum of the input voltages is divided by the number of input voltages applied.

Of course, the above equations apply assuming that the Op Amp is properly nulled, that is, $V_o = 0$ V if all inputs are 0 V. The effect that input bias current has on the output offset and stability can be reduced if we use a resistor R_2 between the noninverting input and ground, see Fig. 8-6, where its value is

$$R_2 = \frac{1}{1/R_F + 1/R_a + 1/R_b + 1/R_c + \cdots + 1/R_n} \qquad (8\text{-}4)$$

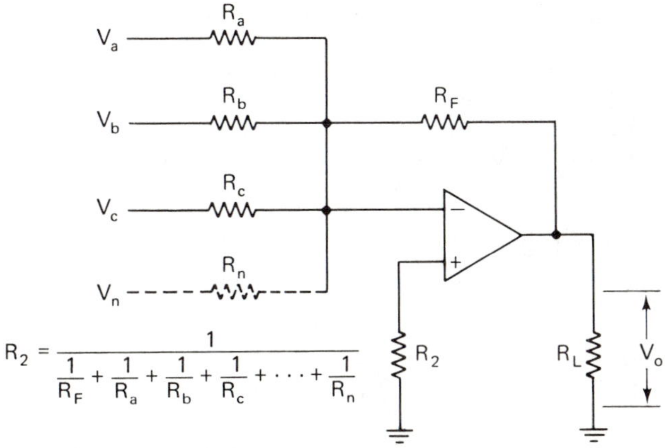

Figure 8-6   Inverting summing or averaging circuit with improved stability.

The input voltage sources and resistors can be connected to the noninverting input of an Op Amp as shown in Fig. 8-7a. These sources and resistors can be replaced with a Norton's equivalent circuit, as shown in Fig. 8-7b, where

$$I_{eq} = \frac{V_a}{R_a} + \frac{V_b}{R_b} + \frac{V_c}{R_c} + \cdots + \frac{V_n}{R_n} \tag{8-5}$$

and

$$R_{eq} = \frac{1}{1/R_a + 1/R_b + 1/R_c + \cdots + 1/R_n} \tag{8-6}$$

Since the resistance looking into the noninverting input is very large, practically all of the current I_{eq} is forced through R_{eq}, resulting in a voltage at the noninverting input 2, that is

$$V_2 = R_{eq}I_{eq} \tag{8-7}$$

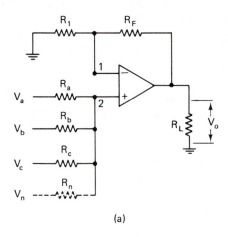

(a)

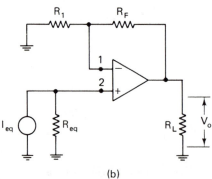

(b)

Figure 8-7 Noninverting averaging circuit.

If all of the input resistors are equal, voltage V_2 is the average of the input voltages. The resulting output voltage is

$$V_o \cong \left(\frac{R_F}{R_1} + 1\right) V_2 \tag{8-8}$$

Example 8-1

Determine the value and polarity of the output voltage in each of the circuits in Fig. 8-8.

Answer.   The circuit shown in Fig. 8-8a is an inverting summing amplifier. Since all of its input resistors are equal, the output voltage can be determined with Eq. (8-2), that is

$$V_o \cong -\frac{30 \text{ k}\Omega}{10 \text{ k}\Omega} (0.2 \text{ V} - 0.15 \text{ V} - 0.35 \text{ V})$$

$$= -3(-0.3 \text{ V}) = +0.9 \text{ V}$$

The circuit in Fig. 8-8b is a noninverting type whose input voltages and resistances are replaced with a Norton's equivalent circuit as shown, where

$$I_{eq} = \frac{-4 \text{ V}}{1 \text{ k}\Omega} + \frac{6 \text{ V}}{1 \text{ k}\Omega} + \frac{5 \text{ V}}{1 \text{ k}\Omega} + \frac{-10 \text{ V}}{1 \text{ k}\Omega} \tag{8-5}$$

$$= -4 \text{ mA} + 6 \text{ mA} + 5 \text{ mA} - 10 \text{ mA} = -3 \text{ mA}$$

and

$$R_{eq} = \frac{1 \text{ k}\Omega}{4} = 250 \text{ }\Omega \tag{8-6}$$

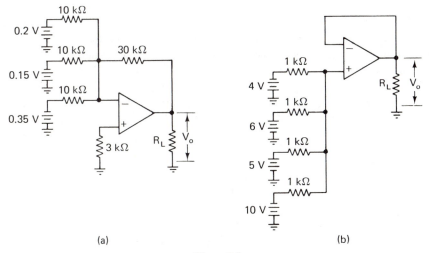

(a)   (b)

Figure 8-8

Therefore, the voltage at the noninverting input

$$V_2 = 250 \ \Omega(-3\text{mA}) = -0.75 \ \text{V} \tag{8-7}$$

which is the average of the input voltages. We can recognize that this Op Amp is connected to work as a voltage follower, which means that the output voltage V_o is the same as the input voltage or -0.75 V in this case.

8.3 THE OP AMP AS AN INTEGRATOR

The Op Amp in Fig. 8-9 is connected to work as an integrator. Its output voltage waveform is the negative integral of the input voltage waveform for properly selected values of R and C. Its operation can be analyzed as follows: Assuming that the Op Amp is ideal, the inverting input 1 and the right side of the input resistor R are at ground potential because the noninverting input is grounded. Therefore, the applied voltage V_s appears across R and the current in this resistor is

$$I \cong \frac{V_s}{R} \tag{8-9}$$

Because of the very large resistance looking into the inverting input, practically all of this current is forced through the capacitor C, which changes the voltage across it. Generally, the current through and the voltage across a capacitor are related by the following equations:

$$i_c = C \frac{dv_c}{dt} \tag{8-10a}$$

$$v_c = \frac{1}{C} \int i_c \ dt \tag{8-10b}$$

Since the left side of the capacitor C in Fig. 8-9 is virtually grounded, then the output voltage V_o is the voltage across the capacitor, and therefore Eq. (8-10b) can be modified to

$$V_o = -\frac{1}{C} \int I \ dt$$

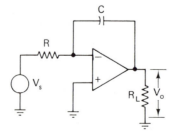

Figure 8-9   Op Amp wired to work as an integrator.

where I is the current being forced through the capacitor. Substituting the right side of Eq. (8-9) into the above results in

$$V_o = -\frac{1}{C} \int \frac{V_s}{R} \, dt$$

or

$$V_o = -\frac{1}{RC} \int V_s \, dt \tag{8-11}$$

where the negative sign represents the phase inverting property of this circuit. If we select the R and C values so that their product is 1, the above equation simplifies to

$$V_o = -\int V_s \, dt \tag{8-12}$$

Example 8-2

If in the circuit of Fig. 8-9, $R = 1$ MΩ, $C = 1$ μF, and V_s has the waveform in Fig. 8-10a, sketch the resulting waveform of the output V_o on the scale in Fig. 8-10b. Assume that the Op Amp is ideal and that the capacitor C is initially uncharged.

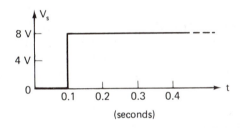

(a)

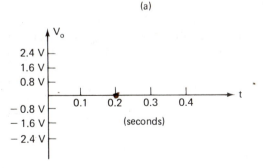

(b)

Figure 8-10

Answer.   The step voltage applied can be expressed as a constant function beginning at $t = 0.1$ s. Since the RC product is 1, Eq. (8-12) applies, and we can show that

$$V_o = - \int V_s \, dt$$

$$= - \int 8 \, dt$$

$$= -8t + V_i = -8t$$

where V_i represents the initial voltage on the capacitor, which in this case is O V. Thus, with a constant voltage function applied to the input of the integrator, the output is a ramp function with a negative slope. Thus the output voltage is a linear function of time, and we can show that since

$$V_{o(t)} = -8t$$

then

$$V_{o(0.1s)} = -8 \times 0.1 = -0.8 \text{ V}$$

and then

$$V_{o(0.2s)} = -8 \times 0.2 = -1.6 \text{ V}$$

etc. Sketching these versus time results in an output waveform as shown in Fig. 8-11. A number of other input and output voltage waveforms of this circuit are shown in Fig. 8-12.

As with the previously discussed Op Amp applications, the integrating circuit of Fig. 8-9 can be made less prone to drift and have less output offset if the proper resistance value is placed between the noninverting input and ground. In this case, this resistance is simply made equal to the input resistance R.

8.4 THE OP AMP AS A DIFFERENTIATOR

The Op Amp can be wired to work as a differentiator. As such, its output is essentially the derivative of the input voltage waveform. The basic differentiator is the circuit shown in Fig. 8-13a. The right side of the input capacitor

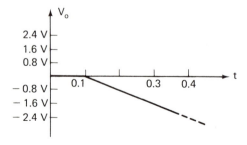

Figure 8-11   Answer to Example 8-3.

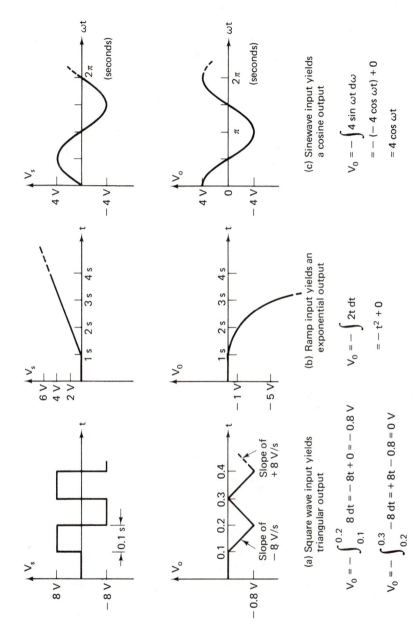

Figure 8-12 Various possible input and output waveforms of the circuit in Figure 8-9 where $R \times C = 1$.

(a) Square wave input yields triangular output

$$V_o = -\int_{0.1}^{0.2} 8\ dt = -8t + 0 = -0.8\ V$$

$$V_o = -\int_{0.2}^{0.3} -8\ dt = +8t - 0.8 = 0\ V$$

(b) Ramp input yields an exponential output

$$V_o = -\int 2t\ dt$$

$$= -t^2 + 0$$

(c) Sinewave input yields a cosine output

$$V_o = -\int 4 \sin \omega t\ d\omega$$

$$= -(-4 \cos \omega t) + 0$$

$$= 4 \cos \omega t$$

147

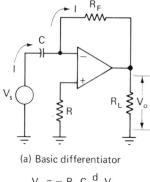

(a) Basic differentiator

$$V_o = - R_F C \frac{d}{dt} V_s$$

$$R = R_F$$

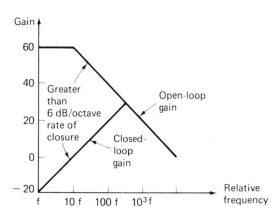

(b) Typical closed-loop and open-loop response curves of the basic differentiator.

Figure 8-13

C is virtually grounded, and therefore the voltage across it is the input voltage V_s. Current I flows to charge or discharge C only when the input voltage changes. Thus, by substituting V_s for v_c and I for i_c in Eq. (8-10a), we can show that

$$I = C \frac{dV_s}{dt} \tag{8-13}$$

Since the resistance looking into the inverting input is very large, any existing input current I is forced up through the feedback resistor R_F. But the left side of R_F is virtually grounded, and therefore the voltage drop across it is the output voltage V_o. By Ohm's law then

$$V_o = R_F I \qquad (8\text{-}14)$$

Rearranging and substituting this last equation into Eq. (8-13), we can show that

$$\frac{V_o}{R_F} = -C \frac{dV_s}{dt}$$

or

$$V_o = -R_F C \frac{d}{dt} (V_s) \qquad (8\text{-}15)$$

Apparently, the output voltage V_o is the derivative of the input voltage V_s times the negative product of R_F and C.

Unfortunately, the circuit in Fig. 8-13 has some practical problems. Because the ratio of the feedback resistance, R_F, to the input capacitor's reactance, X_C, rises with higher frequencies, this circuit's gain increases with frequency. This tends to amplify the high-frequency noise generated in the system, and the resulting output noise can completely override the differentiated signal.

Another problem with this basic differentiator is its tendency to be unstable. In Chapter 6 we learned that, if the closed-loop and open-loop gain vs. frequency curves intersect at a rate of closure greater than 6 dB/octave, the circuit may be unstable. In the case of the circuit in Fig. 8-13a, the input capacitor C causes a low-frequency roll-off that intersects the open-loop curve at a rate of closure greater than 6 dB/octave as shown in Fig. 8-13b. This means that the circuit will probably be unstable. The problem can be solved if we add two components as shown in Fig. 8-14a. Components R_1 and C_1 cause a 6 dB/octave roll-off with decreasing frequencies, while R_F and C_F cause a 6 dB/octave roll-off at higher frequencies, resulting in a closed-loop gain vs. frequency curve as shown in Fig. 8-14b. The components R_1, R_F, and C_F can be selected so that the closed-loop gain vs. frequency curve does *not* intersect with the open-loop curve, thus assuring us of stable operation.

Some input voltages V_s and the resulting differentiated outputs V_o are shown in Fig. 8-15.

8.5 OP AMPS IN VOLTAGE REGULATORS

As mentioned before, three-terminal regulators provide *fixed* regulated dc output voltages inexpensively. In some applications, *variable* output voltages are needed. A three-terminal regulator can be wired as in Fig. 8-16 to

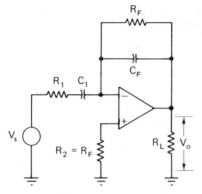

(a) Practical differentiator

$$f_x = \frac{1}{2\pi R_F C_1}$$

$$f_y = \frac{1}{2\pi R_1 C_F}$$

f_z is the frequency at which the open loop-gain is unity

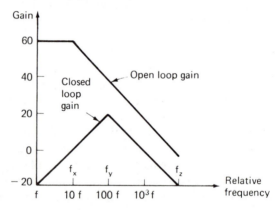

(b) Typical closed-loop and open-loop response curves of the practical differentiator.

Figure 8-14

provide a variable dc output voltage V_{out}. In this case, the LM 340K-5 maintains a regulated 5 V across its pins 2 and 3 and, therefore, across the resistor R_1 too. This circuit's output voltage V_{out} is the sum of the voltage drops across resistors R_1 and R_2. If the resistance of the potentiometer R_2 is changed, the voltage drop across it changes causing a change in the output voltage V_{out}. If R_2 is reduced to 0 Ω, pin 3 is pulled down to ground potential and $V_{out} = 5$ V. If R_2 is increased, V_{out} will increase to a maximum value that is V_{in} minus the dropout voltage of the regulator.

Note in Fig. 8-16 that current I_2 is the sum of the currents I_Q and I_1. Current I_Q, the quiescent current of the regulator, varies slightly if the load current I_L varies. This causes the drop across R_2 to change which in turn causes V_{out} to change. Therefore, the output voltage V_{out} is not as well

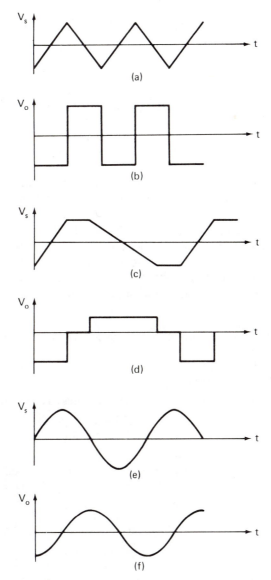

Figure 8-15 Differentiator's input and resulting output waveforms; input waveform (a) causes output waveform (b); input waveform (c) causes output waveform (d); and input waveform (e) causes output waveform (f).

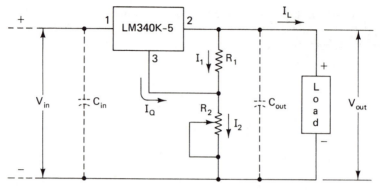

Figure 8-16 Three-terminal regulator wired to work as a variable dc voltage
source.

regulated as the voltage across pins 2 and 3, especially with larger values of
R_2.

A variable and well-regulated output voltage V_{out} can be obtained with
a three-terminal regulator if used with an Op Amp as shown in Fig. 8-17. The
value of V_{out} is selected by adjustment of the POT; resistor R in this case. As
in the previous circuit, variations in the load current I_L causes small
variations in I_Q. In this case, however, I_Q flows through the low output
resistance of the Op Amp instead of the voltage controlling POT. Of course,
then, V_{out} is well regulated. With this circuit, V_{out} can be varied from
approximately 2 V above ground potential to about 2 V less than V_{in}.

Example 8-4

In the circuit of Fig. 8-16, V_{in} varies from 15 V to 20 V, $I_Q = 3$ mA, $R_1 = 2$ kΩ and R_2
is a 1 kΩ POT. What is V_{out}
    (a) when the POT resistance is 1 kΩ, and
    (b) when the POT resistance is 500 Ω?

Answers. Since the voltage across pins 2 and 3 is regulated at 5 V, the current in R_1
is

$$I_1 = \frac{5 \text{ V}}{R_1} = \frac{5 \text{ V}}{2 \text{ k}\Omega} = 2.5 \text{ mA}$$

This adds to I_Q to cause 5.5 mA in R_2. When (a) $R_2 = 1$ kΩ, the resulting drop across
it is 1 kΩ(5.5 mA) or 5.5 V. The sum of the voltages across R_1 and R_2, therefore, is

$$V_{out} = 5 \text{ V} + 5.5 \text{ V} = 10.5 \text{ V}$$

When (b) $R_2 = 0.5$ kΩ, its voltage drop is 0.5 kΩ(5.5 mA) or 2.75 V. In this case,

$$V_{out} = 5 \text{ V} + 2.75 \text{ V} = 7.75 \text{ V}$$

Example 8-5

Referring to the regulator circuit described in the previous example, what maximum
power will the regulator chip dissipate if the load draws up to 200 mA?

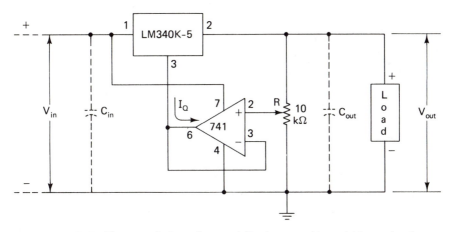

Figure 8-17 Three-terminal regulator and Op Amp provide variable regulated output voltage V_{out}.

Answer.   The maximum voltage across the regulator (pins 1 and 2) occurs when V_{in} is maximum while V_{out} is adjusted to its minimum value. In this case, this maximum voltage is 20 V -- 7.75 V = 12.25 V. Since I_Q and I_1 are negligible compared to the 200 mA maximum load current, the maximum regulator power

$$P \cong 12.25 \text{ V}(200 \text{ mA}) = 2.45 \text{ W}$$

Where the load current is to be in the order of several amperes, the circuit in Fig. 8-18a can be used to provide a regulated output voltage. The transistor Q, sometimes called *the pass element*, drops the difference in the unregulated dc source voltage V_{in} and the load voltage V_{out}. This total current in the emitter of Q is controlled by a much smaller current in its base. Note that the base current is supplied by the Op Amp.*

This circuit can be analyzed as follows: A current maintained through the zener provides a well-regulated voltage at the noninverting input. This zener's voltage V_z causes an Op Amp output voltage V_o that is A_v times larger. The voltage V_o is applied to the base of the transistor Q and forward biases its base-emitter junction. For a silicon transistor, this base-emitter drop V_{BE} is typically about 0.7 V. Therefore, the load voltage V_{out} which is the voltage at the emitter, is less than the Op Amp's output V_o by the base-emitter drop V_{BE}, i.e.,

$$V_{out} = V_o - V_{BE} \qquad (8\text{-}16a)$$

or

$$V_{out} \cong V_o - 0.7 \text{ V} \qquad (8\text{-}16b)$$

* A 50-Ω to 200-Ω resistance is sometimes used in series with the base of Q to protect the circuit with excessively low load resistances.

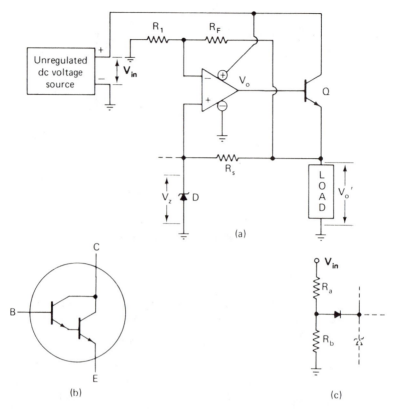

Figure 8-18 (a) Regulated dc source with intermediate current capability; (b) Darlington pair can be used in place of transistor Q for higher load-current capability; (c) voltage divider that can be used to assure starting (cause zener to go into avalanche conduction).

For most practical purposes,

$$V_{out} \cong V_o \tag{8-16c}$$

Since the Op Amp is wired to work as a noninverting amplifier,

$$V_{out} \cong A_v V_z \tag{8-17a}$$

or

$$V_{out} \cong \left(\frac{R_F}{R_1} + 1\right)V_z \tag{8-17b}$$

In the circuit of Fig. 8-18a, the drop across R_s is the difference in the load and zener voltages, $V_{out} - V_z$. Since V_{out} is regulated, this resistor's drop is quite constant, and therefore, so is the current through it and the zener. A constant current through the zener will cause a very constant voltage drop across it regardless of its dynamic impedance Z_{zt} and this gives this circuit its good voltage-regulating capability.

This circuit's components can be selected as follows: Resistance R_s is found by the Ohm's law equation

$$R_s = \frac{V_{out} - V_z}{I_z}$$

where I_z is the zener current and is selected to be between I_{zm} and I_{zt}; see Table 8-1. The minimum value of the unregulated voltage V_{in} must be larger than the required load voltage V_{out}. To put it another way, the regulated load voltage V_{out} must be less than the minimum V_{in}. The difference in V_{in} and V_{out} is dropped across the transistor Q. This transistor is selected so that its maximum collector current capability exceeds the maximum current we expect the load to draw, plus the zener current I_z, that is,

$$I_{C(max)} > I_{L(max)} + I_z \qquad (8\text{-}18)$$

where $I_{C(max)}$ is the maximum collector current, and
$\quad I_{L(max)}$ is the maximum load current.
Also, the transistor's power rating must exceed

$$P_{C(max)} = (V_{in(max)} - V_{out})I_{C(max)} \qquad (8\text{-}19)$$

The zener's power rating must exceed its power dissipation as determined with the equation

$$P_z = V_z I_z \qquad (8\text{-}20)$$

It is possible that when this circuit is initially turned on, the transistor Q may not turn on hard enough to provide a load voltage V_{out} that is greater than V_z. In such cases, the zener does not go into avalanche conduction, and the circuit will not regulate the load voltage. A starting voltage can be used across the zener, provided by a voltage divider such as that in Fig. 8-18c. The resistors R_a and R_b are selected so that the drop across R_b exceeds the

TABLE 8-1.   1-Watt Zener Diodes

JEDEC Type No.	Nominal Zener Voltage V_z @ I_{zt} Volts	Test Current I_{zt} mA	Max Zener Impedance		Max DC Zener Current I_{zm} mA
			Z_{zt} @ I_{zt} Ohms	Z_{zk} @ I_{zk} = 1.0 mA Ohms	
IN3821	3.3	76	10	400	276
IN3822	3.6	69	10	400	252
IN3823	3.9	64	9	400	238
IN3824	4.3	58	9	400	213
IN3825	4.7	53	8	500	194
IN3826	5.1	49	7	550	178
IN3827	5.6	45	5	600	162
IN3828	6.2	41	2	700	146
IN3829	6.8	37	1.5	500	133
IN3830	7.5	34	1.5	250	121

zener voltage V_z, that is, so that

$$\frac{V_{in(min)}(R_b)}{R_a + R_b} > V_z \qquad (8\text{-}21)$$

The regulated output voltage V_{out} from the circuit in Fig. 8-18 can be made variable, within limits, by replacing R_1 and R_F with a potentiometer. The wiper of the potentiometer is connected to the inverting input of the Op Amp.

The load current capability of this regulator can be increased up to several amperes by using a power *darlington* pair in place of the transistor Q. Typically, the pass transistor Q or darlington pair is mounted on a heat sink.

Example 8-6

In the circuit of Fig. 8-18a, suppose that V_{in} varies between 20 V and 24 V, V_{out} is to be regulated at 18.1 V using a 1N3825-type zener, and the load draws from zero to 2 A. Select the proper ratio of R_F/R_1, the approximate ratio R_a/R_b, and the resistance of R_s. What maximum possible current does the transistor Q draw and what maximum power does it dissipate? How much power does the zener dissipate?

Answer. Given the required regulated output V_{out} of 18.1 V and $V_z = 4.7$ V; the zener voltage of the 1N3825 shown in Table 8-1, we can determine the ratio R_F/R_1 by rearranging Eq. (8-17a). Since

$$A_v \cong \frac{V_{out}}{V_z}$$

in this case, we need an

$$A_v \cong \frac{18.1 \text{ V}}{4.7 \text{ V}} = 3.85$$

And since

$$A_v \cong \frac{R_F}{R_1} + 1$$

then

$$\frac{R_F}{R_1} \cong A_v - 1$$

and therefore

$$\frac{R_F}{R_1} \cong 2.85$$

A starting voltage can be provided with a voltage divider such as that in Fig. 8-18c. In this case,

$$\frac{V_{in(min)}(R_b)}{R_a + R_b} > V_z = 4.7 \text{ V}$$

and therefore

$$\frac{20 \text{ V}(R_b)}{R_a + R_b} > 4.7 \text{ V}$$

from which we can show that, in order to have reliable starting, we must use a ratio $R_a/R_b < 3.26$.

Since Table 8-1 shows that the test current of the 1N3825 is 53 mA, we can use an

$$R_s = \frac{V_{out} - V_z}{I_z} = \frac{18.1 \text{ V} - 4.7 \text{ V}}{53 \text{ mA}} \cong 250 \text{ }\Omega$$

The transistor will have to carry as much as

$$I_{C(max)} = I_{L(max)} + I_z \cong 2.053 \text{ A}$$

The maximum transistor power dissipation is

$$P_{C(max)} = (24 \text{ V} - 18.1 \text{ V}) 2.053 \text{ A} \cong 12.1 \text{ W}$$

And the zener's power dissipation is

$$P_z = 4.7 \text{ V}(53 \text{ mA}) \cong 250 \text{ mW}$$

REVIEW QUESTIONS

8-1. Generally, what is the meaning of the term *linear application* of an Op Amp?

8-2. Since Op Amps are capable of amplifying dc and ac voltages, why should we ever consider using coupling capacitors between Op Amp stages in ac amplifier systems?

8-3. What is the meaning of and the reason for bootstrapping the input of an amplifier?

8-4. If an Op Amp wired as an integrator has a squarewave input voltage applied, what output waveform can we expect?

8-5. If an Op Amp wired to work as a differentiator has a sawtooth input voltage applied, what output voltage waveform can we expect?

8-6. What advantage does the Op Amp give to the circuit of Fig. 8-17 compared to the circuit of Fig. 8-16?

8-7. Why would a circuit like that of Fig. 8-18 be used instead of the circuit of Fig. 8-17?

8-8. In the circuit of Fig. 8-19, what is the purpose of the 240-kΩ resistor?

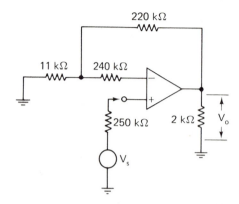

Figure 8-19

PROBLEMS

8-1. What is the gain V_o/V_s of the circuit in Fig. 8-19?

8-2. In the circuit of Fig. 8-19, if the 11-kΩ resistor is replaced with an 82-kΩ resistor, (a) what is the gain V_o/V_s of the circuit, and (b) with what value should the 240-kΩ resistor be replaced if the circuit stability is to remain close to what it was?

8-3. In the circuit of Fig. 8-20, (a) what is the gain V_o/V_s of the circuit, and (b) what value of resistance R should we use to minimize instability and output offset voltage?

8-4. If the Op Amp in the circuit of Fig. 8-20 is a type 741 and $R = 1$ MΩ, roughly what maximum output offset voltage, caused by bias currents, could we expect at room temperature?

8-5. In the circuit of Fig. 8-21, what total gain V_o/V_s can we expect at frequencies at which the reactances of all capacitors are negligible if the potentiometer is adjusted to 0 Ω?

8-6. At frequencies that cause the reactances of the capacitors to be negligible in the circuit of Fig. 8-21, what gain V_o/V_s can we expect if the potentiometer is adjusted to maximum resistance?

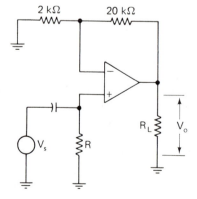

Figure 8-20

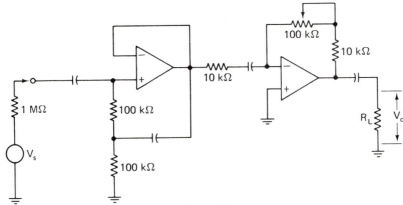

Figure 8-21

8-7. In the circuit of Fig. 8-16, V_{in} varies from 15 V to 20 V, $I_Q = 3$ mA, $R_1 = 2$ kΩ and R_2 is a 1 kΩ POT. What is V_{out} when (a) the POT resistance is 800 Ω, and when (b) the POT resistance is 100 Ω?

8-8. In the circuit of Fig. 8-16, V_{in} varies from 20 V to 24 V, $I_Q = 4$ mA, $R_1 = 3.3$ kΩ, and R_2 is a 2-kΩ POT. What is V_{out} when (a) the POT resistance is 2 kΩ, and when (b) the POT resistance is 0 Ω?

8-9. Referring to the circuit described in Prob. 8-7, if a load current change causes I_Q to change by 0.5 mA, what is the resulting change in the output voltage V_{out} when the POT resistance is 800 Ω?

8-10. Referring to the circuit described in Prob. 8-8, if a load current change causes I_Q to change by 0.5 mA, what is the resulting change in the output V_{out} when the POT resistance is 2 kΩ?

8-11. If, in the circuit of Fig. 8-17, a 3.3-kΩ resistor is placed in series above the 10-kΩ POT while a 1.2-kΩ resistor is placed in series below the 10-kΩ POT, over what range can V_{out} be adjusted, theoretically? *Hint:* The voltage across pin 2 of the regulator and pin 2 of the Op Amp is a regulated 5 V. Also, pins 2 and 6 of the Op Amp are at the same potential.

8-12. If, in the circuit of Fig. 8-17, a 10-kΩ resistor is placed in series above the 10-kΩ POT and a 2-kΩ resistor is placed in series below the 10-kΩ POT, over what range can V_{out} be adjusted, theoretically?

8-13. Referring to the circuit of Fig. 8-18, V_{in} varies from 9 V to 12 V and V_{out} is to be regulated at about 5 V. The load draws up to 3 A. Use a 1N3821 zener and an $I_z = 75$ mA. Select the proper ratios of resistors R_F/R_1, of R_a/R_b and the value of R_s. With your component values, what maximum power will the transistor and the zener be required to dissipate?

8-14. In the circuit of Fig. 8-22, if $R_1 = R_2 = R_3 = 1$ kΩ and if the Op Amp was initially nulled, (a) what is its output voltage V_o, and (b) what value of R_4 should we use to minimize drift?

8-15. If $R_1 = 1$ kΩ, $R_2 = 2$ kΩ, and $R_3 = 4$ kΩ in the circuit of Fig. 8-22, (a) what is its output voltage V_o, and (b) what value of R_4 will minimize drift? Assume that the Op Amp was initially nulled.

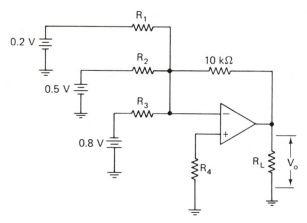

Figure 8-22

8-16. The three input resistors R are to be equal in the circuit of Fig. 8-23. (a) In order to minimize drift, what value should each of these input resistors be? (b) What is the voltage at the noninverting input, and (c) what is the voltage V_o? Assume that the Op Amp was nulled.

8-17. In the circuit of Fig. 8-9, if $R = 100$ kΩ, $C = 100$ μF, and the applied voltage V_s has the waveform shown in Fig. 8-10. What is the output voltage at (a) $t = 1.1$ second and at (b) $t = 5.1$ seconds. Assume that initially the Op Amp was nulled and that the capacitor's voltage was zero.

8-18. In the circuit described in Prob. 8-17, how long will it take for the amplifier to saturate after the 8-V step voltage is applied, if the output saturation voltages are $+ 16$ V and $- 16$ V?

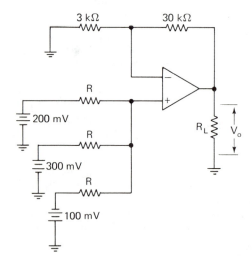

Figure 8-23

9

ACTIVE FILTERS

Electrical filters were in use long before IC Op Amps were available. Typically they were constructed with passive components; resistors, capacitors and inductors, and were in fact called passive filters. With low cost and reliable amplifying (active) devices available, such as Op Amps, filter designs are now mainly active types. For most applications, active filters are easy to design and they eliminate the need for high cost inductors.

In this chapter we will find that active filters are classified in four catagories. These are

1. Low pass filters
2. High pass filters
3. Bandpass filters, and
4. Bandstop filters

Low pass filters allow low frequencies to pass while rejecting higher values. High pass filters reject low frequencies but pass the highs. Bandpass filters serve to amplify or pass a narrow range of frequencies while attenuating or rejecting all others. And, the bandstop filters are used where a narrow range of frequencies is to be rejected while all others are to pass through.

9.1 BASIC RC FILTERS

The series RC circuit of Fig. 9-1 is a simple *low pass* (LP) filter. Its frequency response is shown in part (b) of this figure. At dc and relatively low frequencies (lower than f_c), the output V_{out} is about at the same amplitude as the input V_{in}. Higher frequencies of V_{in}, however, cause the reactance of the capacitor C to decrease. This causes an increased voltage drop across the resistance R and a corresponding decrease in the output voltage V_{out}. Generally,

$$V_{out} = \frac{V_{in}(-jX_C)}{R - jX_C} \tag{9-1a}$$

where

$$-jX_C = \frac{-j}{2\pi f C} = \frac{-1}{j\omega C} \tag{9-2}$$

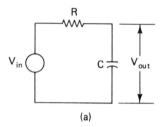

(a)

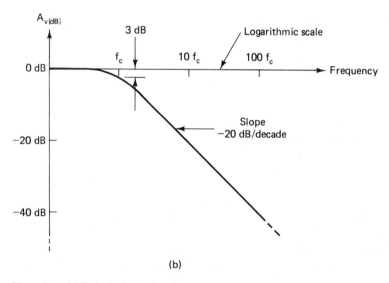

(b)

Figure 9-1   (a) Series RC circuit as low pass filter and (b) its frequency response.

or, neglecting phase shift

$$V_{\text{out}} = \frac{V_{\text{in}} X_C}{\sqrt{R^2 + X_C^2}} \tag{9-1b}$$

Gain A_v is the ratio of the output V_{out} to the input V_{in}. Therefore,

$$A_v = \frac{V_{\text{out}}}{V_{\text{in}}} = \frac{-jX_C}{R - jX_C} \tag{9-3a}$$

or

$$|A_v| = \left| \frac{V_{\text{out}}}{V_{\text{in}}} \right| = \frac{X_C}{\sqrt{R^2 + X_C^2}} \tag{9-3b}$$

The phase shift of V_{out} with respect to V_{in} is

$$\phi = -\arctan \left(\frac{R}{X_C} \right) = -\arccos \left(\frac{V_{\text{out}}}{V_{\text{in}}} \right)$$

In decibels, the gain

$$A_{v(\text{dB})} = 20 \log \left| \frac{V_{\text{out}}}{V_{\text{in}}} \right| = 20 \log |A_v| \tag{9-4}$$

The frequency at which the response curve starts to bend (breaks) is called the *cutoff frequency* f_c, *critical frequency*, or *break frequency*. The gain $A_{v(\text{dB})}$ is 3 dB below its maximum value at f_c [see (b) of Fig. 9-1]. Note that at frequencies higher than f_c, the gain $A_{v(\text{dB})}$ rolls off at a 20 dB/decade rate.

The use of resistance and capacitance to filter (remove) higher frequencies is not new to us. We already know that filter capacitors are placed across the outputs of dc power supplies to filter out the ripple frequencies and transient pulses. Even the 741 Op Amp has an internal 30 pF capacitor that, combined with internal resistance of its circuitry, causes the -20 dB/decade to roll off of its open loop vs. frequency response curve (examine the circuit and curve of Appendix I).

The RC circuit of Fig. 9-2 works as a *high pass* (HP) filter. At higher frequencies, the reactance of the capacitor C is negligible, causing V_{in} to appear across the resistor R; that is, $V_{\text{out}} \cong V_{\text{in}}$. At lower frequencies, the increased voltage drop across C causes V_{out} to decrease, as shown in (b) of Fig. 9-2. In this case,

$$A_v = \frac{V_{\text{out}}}{V_{\text{in}}} = \frac{R}{R - jX_C} \tag{9-5a}$$

or

$$|A_v| = \left| \frac{V_{\text{out}}}{V_{\text{in}}} \right| = \frac{R}{\sqrt{R^2 + X_C^2}} \tag{9-5b}$$

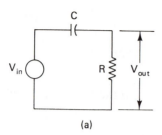

(a)

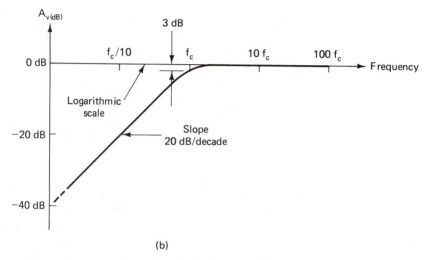

(b)

Figure 9-2   (a) Series RC circuit as high pass filter and (b) its frequency response.

The phase shift of V_{out} with respect to V_{in} is

$$\phi = \arctan\left(\frac{X_C}{R}\right) = \arccos\left(\frac{V_{\text{out}}}{V_{\text{in}}}\right)$$

As with the LP filter, $A_{v(\text{dB})}$ is 3 dB below its maximum value at the cutoff frequency f_c. As shown, at frequencies lower than f_c, the gain $A_{v(\text{dB})}$ vs. frequency curve has a slope of 20 dB/decade.

In either circuit of Fig. 9-1 or Fig. 9-2, the reactance of the capacitor C is equal to the resistance R at the cutoff frequency f_c. Therefore, since

$$X_C = \frac{1}{2\pi f C}$$

then

$$f = \frac{1}{2\pi X_C C}$$

And at f_c, where $X_C = R$,

$$f_c = \frac{1}{2\pi RC} \qquad (9\text{-}6)$$

This equation enables us to determine quickly the cutoff frequency of a simple RC filter.

9.2 FIRST-ORDER ACTIVE FILTERS

The simple RC filters of Figs. 9-1 and 9-2 cannot provide consistent filtering if V_{out} must be applied across a low or varying load resistance. An Op Amp can serve to buffer (isolate) the load from the filter, as shown in Fig. 9-3. Both circuits are first-order active filters. Generally, active filters use components or devices that are able to amplify, such as transistors or Op

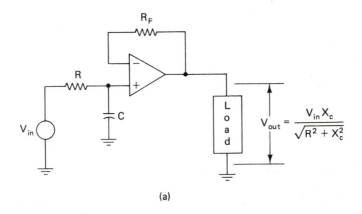

(a)

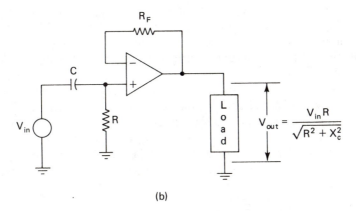

(b)

Figure 9-3   First-order active filters: (a) low pass and (b) high pass.

Amps. Passive filters consist of passive components only: resistors, capacitors, and/or inductors. A first-order filter has one resistor R and one capacitor C; a second-order filter has two resistors and two capacitors; a third-order filter has three and three, and so on.

Both Op Amps of Fig. 9-3 are wired to work as voltage followers. A resistance $R_F = R$ is used if drifting bias currents cause excessive drift in the quiescent value of V_{out}. Otherwise, R_F is replaced with a short (0 Ω).

Example 9-1

If $R = R_F = 20$ kΩ and $C = 800$ pF in the circuit (a) of Fig. 9-3, find the cutoff frequency f_c. Solve for the gain $A_{v(dB)}$ at (a) $f_c/100$, (b) $f_c/10$, (c) f_c, (d) $10 f_c$ and (e) 100 f_c.

Answers. By Eq. 9-6,

$$f_c = \frac{1}{2\pi(20 \text{ kΩ})800 \text{ pF}} \cong 9947.2 \text{ Hz; call it 10 kHz in round numbers.}$$

Now by Eqs. 9-2, 9-3b, and 9-4, we find that
   (a) at $f = 100$ Hz, $A_{v(dB)} \cong -4.4 \times 10^{-4}$dB $\cong$ OdB
   (b) at $f = 1$ kHz, $A_{v(dB)} \cong -4.4 \times 10^{-2}$dB $\cong$ OdB
   (c) at $f = 10$ kHz, $A_{v(dB)} \cong -3$ dB
   (d) at $f = 100$ kHz, $A_{v(dB)} \cong -20$ dB
   (e) at $f = 1$ MHz, $A_{v(dB)} \cong -40$ dB
If you plot these $A_{v(dB)}$ vs. f values on semilog paper, you'll get a curve like (b) of Fig. 9-1.

9.3 NTH ORDER ACTIVE FILTERS

To generalize, we can say that an LP filter passes frequencies below the cutoff frequency f_c and attenuates frequencies above f_c. In most applications, the −20 dB/decade roll-off of $A_{v(dB)}$ is inadequate. Ideally, an LP filter's $A_{v(dB)}$ vs. f curve should roll off (decrease) vertically above f_c (see curve 5 in Fig. 9-4). An ideal curve is difficult to achieve but improvements are quite practical. An improved curve has a roll-off that is steeper than −20 dB/decade.

By cascading two first-order LP filters, we can have a second-order LP filter. Three first-order filters cascaded can yield a third-order filter, etc. As shown in Fig. 9-4, a second-order LP filter has a −40 dB/decade roll-off, and a third-order LP filter has a −60 dB/decade roll-off. Obviously, each additional first-order filter placed in cascade increases the roll-off by another −20 dB/decade. However, since each Op Amp adds noise, the number that can be cascaded is limited. Later we will see how to build a second-order filter with only one Op Amp.

Figure 9-5 shows three first-order LP active filters cascaded. Let's say,

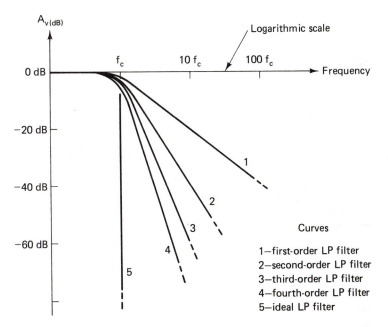

Figure 9-4 $A_{v(dB)}$ vs. frequency curves if LP filters: first order, second order, third order, etc.

for example, that the first stage has a cutoff frequency of f_{c1}, the second stage has a higher cutoff of f_{c2}, and that the third stage has an even higher cutoff of f_{c3}. Their individual $A_{v(dB)}$ vs. f curves are shown in (a)–(c) of Fig. 9-6. The curve for the circuit as a whole is shown in (d) of this figure. This overall $A_{v(dB)}$ vs. f curve is the result of superimposing the individual stages' curves. Recall that the uncompensated Op Amp has a similar open loop vs. frequency curve which is caused by the multiple stages within the Op Amp (see Fig. 6-3).

Note in Fig. 9-6 that at frequencies higher than f_{c3}, the gain of each stage is decreasing at -20 dB/decade, causing an overall $A_{v(dB)}$ roll-off of -60 dB/decade. If all three first-order filters in Fig. 9-5 are designed to have

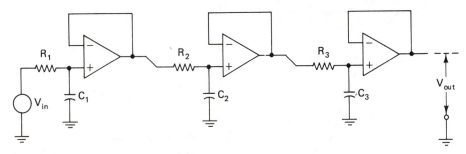

Figure 9-5   Three first-order LP filters casaded.

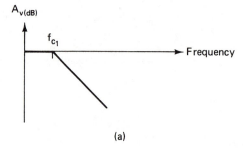

(a)

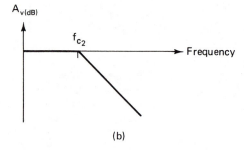

(b)

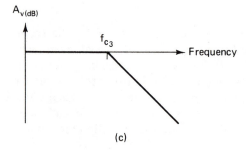

(c)

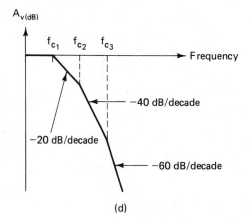

(d)

Figure 9-6 (a) through (c) are $A_{v(dB)}$ vs. frequency curves of three individual LP filters, and (d) their curve when casaded.

the same cutoff frequency f_c, the circut as a whole is a third-order filter. Its $A_{v(dB)}$ vs. f curve rolls off at 60 dB/decade at frequencies above f_c (see curve 3 of Fig. 9-4.)

Example 9-2

Referring to the circuit of Fig. 9-5, if $R_1 = 160$ kΩ, $C_1 = 200$ pF, $R_2 = 270$ kΩ, $C_2 = 50$ pF, $R_3 = 2.7$ kΩ, and $C_3 = 4000$ pF, find the cutoff frequency of each stage. Sketch the overall circuit $A_{v(dB)}$ vs. f curve.

Answers.   Using Eq. 9-6, we find that for the first stage, the cutoff frequency

$$f_{c1} = \frac{1}{2\ \pi(160\text{ k}\Omega)200\text{ pF}} \cong 4.97\text{ kHz}$$

For the second stage,

$$f_{c2} = \frac{1}{2\ \pi\ (270\text{ k}\Omega)50\text{ pF}} \cong 11.8\text{ kHz}$$

And for the third stage

$$f_{c3} = \frac{1}{2\ \pi\ (2.7\text{ k}\Omega)4000\text{ pF}} \cong 14.7\text{ kHz}$$

The curve should look like the one in (d) of Fig. 9-6. At frequencies lower than f_{c1}, the curve is flat. Approaching higher frequencies, the curve breaks at about 4.97 kHz and rolls off at a 20 dB/decade rate. Approaching still higher frequencies, the curve breaks again and, at about 11.8 kHz, rolls off at a 40 dB/decade rate. Moving further to the right on the frequency axis, the curve breaks once more at about 14.7 kHz and rolls off at a 60 dB/decade rate.

Example 9-3

Referring again to the circuit of Fig. 9-5, if $R_1 = R_2 = R_3 = 470$ kΩ and $C_1 = C_2 = C_3 = 330$ pF, find the cutoff frequency of each stage and sketch the overall (total) $A_{v(dB)}$ vs. f response curve.

Answers.   Each stage has the same cutoff frequency. By Eq. (9-6), we find it is

$$f_c = \frac{1}{2\ \pi\ (470\text{ k}\Omega)330\text{ pF}} \cong 1026\text{ Hz}$$

The curve is flat at frequencies well below f_c. Approaching higher frequencies, it breaks at about 1 kHz and rolls off at a 60 dB/decade rate (see curve 3 of Fig. 9-4).

If first-order *high pass* (HP) filters are cascaded, as in Fig. 9-7, they can work as a higher order HP filter. If the resistors and capacitors are selected so that both filters attenuate frequencies below the same cutoff frequency f_c, then each stage has an $A_{v(dB)}$ vs. f curve like #1 in Fig. 9-8. The overall $A_{v(dB)}$ vs. f response of both stages is illustrated by curve #2. Add a third

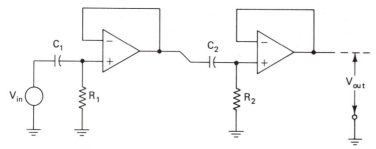

Figure 9-7 First-order high pass filters casaded.

stage and curve #3 represents the resulting $A_{v(dB)}$ response, etc. Of course, the high frequencies we can expect to pass through an active HP filter are limited by the bandwidth of the Op Amp.

Example 9-4

Referring to the circuit of Fig. 9-7, if $R_1 = R_2 = 100\ k\Omega$ and $C_1 = C_2 = 0.008\ \mu F$, find f_c. Determine the overall gain $A_{v(dB)}$ of this two-stage filter at each of the following frequencies: (a) $f_c/100$, (b) $f_c/10$, (c) f_c, (d) $10\ f_c$, and (e) $100\ f_c$.

Answers. By Eq. 9-6

$$f_c = \frac{1}{2\ \pi\ (100\ k\Omega)0.008\ \mu F} = 198.9\ Hz;\ \text{call it 200 Hz in round numbers.}$$

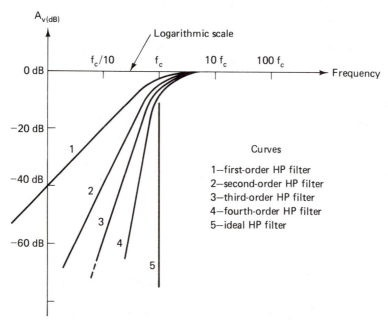

Figure 9-8 $A_{v(dB)}$ vs. f curves of HP filters; first order, second order, third order, etc.

Now by Eq. 9-2, 9-5b, and 9-4, we find that

(a) at $f = 2$ Hz, $A_{v(dB)} \cong -40$ dB for each stage. For two stages in cascade, the total $A_{v(dB)} \cong 2(-40$ dB$) = -80$ dB.

(b) At $f = 20$ Hz, $A_{v(dB)} \cong -20$ dB for each stage or a total $A_{v(dB)} \cong -40$ dB.

(c) At $f = 200$ Hz, $A_{v(dB)} \cong -3$ dB for each stage or a total $A_{v(dB)} \cong 2(-3$dB$) = -6$ dB.

(d) At $f = 2$ kHz, $A_{v(dB)} \cong -0.043$ dB $\cong 0$ dB for each and both stages.

(e) At $f = 20$ kHz, $A_{v(dB)} \cong -0.00043$ dB $\cong 0$ dB for each and both stages.

Plot these on semilog paper and you'll get curve #2 of Fig. 9-8.

9.4 TYPICAL SECOND-ORDER LOW PASS ACTIVE FILTER

The circuit of Fig. 9-9 shows how a single Op Amp can be used in a second-order LP active filter. With proper selection of components, this circuit has a variety of possible $A_{v(dB)}$ vs. f characteristics (see Fig. 9-10). Note that all curves roll-off at -40 dB/decade at frequencies well above the cutoff frequency f_c. As with first-order filters, f_c is determined by the resistor and capacitor values.

The curves of Fig. 9-10 are identified as Bessel, Butterworth, or Chebyshev. These are the names of persons whose work contributed to filter technology. Depending on which $A_{v(dB)}$ vs. f curve a given filter has, it is referred to by one of these names.

The Bessel, Butterworth, and Chebyshev filters differ in their damping coefficients d. The Bessel filter has the most damped characteristic, whereas the Chebyshev types are among the least damped. Bessel and Butterworth filters are used where flat low frequency (below f_c) and good transient responses are needed. Generally, the Chebyshev filters have better (steeper) initial roll-off but are prone to ringing and overshoot in response to transient or pulsed input signals. The Bessel and Butterworth filters are known for

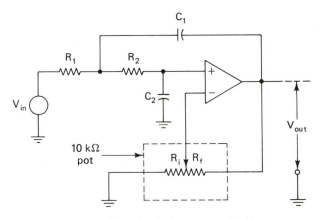

Figure 9-9   Second-order low pass active filter.

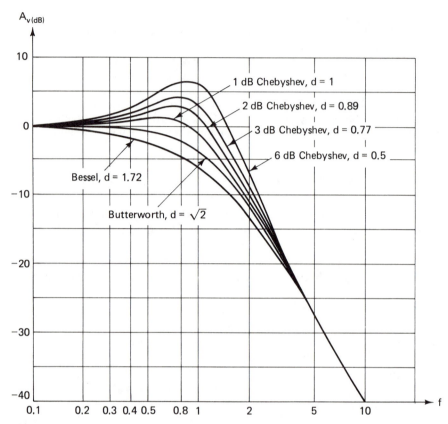

Figure 9-10 Second-order gain $A_{v(dB)}$ vs. frequency f response curves with various dampling coefficients d.

their more linear phase* vs. frequency responses compared to the Chebyshev types.

Each filter has a damping coefficient. As shown in Fig. 9-10, $d = 1.72$ for the Bessel filter, $\sqrt{2}$ for the Butterworth, and so on. The decibel designations on the Chebyshev curves refer to the *ripple channels* of the filters. Figure 9-11 illustrates the ripple channel for 3-dB Chebyshev filters of various orders. In most practical cases, the ripple channels are limited to 3 dB.

The damping coefficient for a second-order filter of Fig. 9-9 can be determined with the equation

$$d = \frac{R_1C_2 + R_2C_2 + R_1C_1(1 - A_v)}{\sqrt{R_1R_2C_1C_2}} \qquad (9\text{-}7)$$

where A_v is the closed-loop gain of the Op Amp

---

* See Appendix S for equations that determine phase shift caused by second-order active filters.

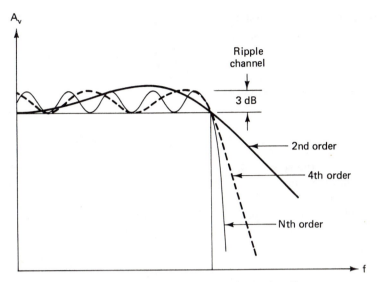

Figure 9-11   Characteristics of 3-dB Chebyshev filters.

The cutoff frequency, in rad/s, is

$$\omega_c = \frac{1}{\sqrt{R_1 R_2 C_1 C_2}} \tag{9-8a}$$

or in Hz

$$f_c = \frac{1}{2\pi\sqrt{R_1 R_2 C_1 C_2}} \tag{9-8b}$$

Obviously, the component values determine the damping coefficient and cutoff frequency, which in turn dictate the shape of the $A_{v(dB)}$ vs. f curve. Change any one value and both d and f_c change also. Filter designers have elaborate tables available that simplify component selection for active filters. In our case, we will simplify matters by limiting the Op Amp's closed-loop gain to unity; that is, $A_v = 1$. In the circuit of Fig. 9-9 then, we can adjust the POT so that $R_f = 0\ \Omega$, or remove the POT and simply wire the Op Amp as a voltage follower. The POT is often used as a trim control. We will also make $R_1 = R_2$. With $A_v = 1$ and $R_1 = R_2 = R$, Eq. (9-8b) becomes

$$f_c = \frac{1}{2\pi R\sqrt{C_1 C_2}}$$

and can be rearranged to

$$R = \frac{1}{2\pi f_c \sqrt{C_1 C_2}} \tag{9-9a}$$

Similarly, Eq. (9-7) becomes

$$d = 2 \sqrt{\frac{C_2}{C_1}}$$

and when rearranged, becomes

$$\frac{C_2}{C_1} = \frac{d^2}{4} \tag{9-10}$$

Thus, knowing the type of filter needed, Bessel, Butterworth, or Chebyshev, we determine the dampling coefficient from the curves of Fig. 9-10. Equation (9-10) then enables us to select C_1 and C_2 from available capacitor values and types.* After C_1 and C_2 are selected use Eq. (9-9a) to find the R values in the *Butterworth* filter. For Bessel and Chebyshev types, Eq. (9-9a) is modified to

$$R = \frac{K}{2 \pi f_x \sqrt{C_1 C_2}} \tag{9-9b}$$

where K is a correction factor determined from Table 9-1 and $f_x = f_c$ or f_p depending on the type of filter being designed. The correction factor is used because f_c was initially defined as the 3-dB down frequency. However, note in Fig. 9-10, that only the Butterworth filter's curve is at -3 dB at f_c. With the correction factors K listed in Table 9-1, f_x is the 3 dB down frequency for the Bessel filter whereas f_x is the peak frequency for Chebyshev types. At the peak frequency f_p, the gain $A_{v(dB)}$ reaches its peak. This is just before the $A_{v(dB)}$ vs. f curve enters the -40 db/decade roll-off.

Example 9-5

We need a second-order LP filter with a $A_{v(dB)}$ vs. f curve that rolls off at -40 dB/decade at frequencies above 4 kHz. Select components for (a) a Butterworth filter and let $f_c = 4$ kHz. Also, select components for (b) a 3-dB Chebyshev filter and let $f_p = 4$ kHz.

TABLE 9-1. Second-Order Correction Factors

Filter	Correction factor
Bessel	0.79 for f_c
Butterworth	1 for f_c
1-dB Chebyshev	0.67 for f_p
2-dB Chebyshev	0.8 for f_p
3-dB Chebyshev	0.84 for f_p

* Capacitors with low leakage and good temperature stability should be used, for example, metalized polycarbonate, or silvered mica.

Answers.   (a) For the Butterworth filter, $d = \sqrt{2} \cong 1.414$. By Eq. 9-10,

$$\frac{C_2}{C_1} = \frac{1.414^2}{4} = 0.5$$

If we choose $C_1 = 0.02 \ \mu F$, then $C_2 = 0.01 \ \mu F$. By Eq. (9-9a),

$$R = \frac{1}{2 \ \pi \ f_c \sqrt{C_1 C_2}} = \frac{1}{2 \ \pi \ (4 \ \text{kHz}) \sqrt{0.02 \ \mu F (0.01 \ \mu F)}} = 2813.5 \ \Omega$$

Use

$$R = 2.7 \ \text{k}\Omega^*$$

(b) For the 3-dB Chebyshev type, $d = 0.77$ and

$$\frac{C_2}{C_1} = \frac{0.77^2}{4} \cong 0.148$$

If we choose $C_1 = 0.02 \ \mu F$, then $C_2 = 0.148 C_1 = 0.0029 \ \mu F$; use $0.003 \ \mu F$. In this case, by Eq. (9-9b)

$$R = \frac{0.84}{2 \pi (4 \ \text{kHz}) \sqrt{0.02 \ \mu F (0.003 \ \mu F)}} = 4.315 \ \text{k}\Omega; \text{ use } 4.3 \ \text{k}\Omega^\dagger$$

9.5 EQUAL COMPONENT LOW PASS ACTIVE FILTER

If for a damping coefficient that we need we find it difficult to find available capacitors that satisfy Eq. (9-10), we can take the following approach: Let $R_1 = R_2 = R$ and $C_1 = C_2 = C$ in the circuit of Fig. 9-9. In this case, Eq. (9-7) simplifies to

$$d = 3 - A_v \qquad (9\text{-}11\text{a})$$

and therefore,

$$A_v = 3 - d \qquad (9\text{-}11\text{b})$$

Also, Eq. (9-8b) becomes

$$f_c = \frac{1}{2 \ \pi \ RC}$$

and, therefore, for the Butterworth filter,

$$R = \frac{1}{2 \ \pi \ f_c C} \qquad (9\text{-}12\text{a})$$

---

* Precision resistors can be used to obtain resistances closer to calculated values.

$\dagger$ See Appendix O for **BASIC** programs that will tabulate $A_{v(\text{dB})}$ vs. frequency characteristics for second-order active filters.

For Bessel and Chebyshev filters,

$$R = \frac{K}{2\pi f_x C} \tag{9-12b}$$

where K is a correction factor obtained from Table 9-1, and as before,

$$f_x = f_c \text{ or } f_p$$

Equation (9-11b) shows that the Op Amp's voltage gain is *not* unity for the equal-component LP filter. In this case, we use and adjust the potentiometer so that $A_v = 3 - d$ or we can select appropriate fixed resistors R_f and R_i.

Example 9-6

Design a 3-dB Chebyshev LP filter with a peak frequency $f_p = 4$ kHz. Use the equal-components approach.

Answer.   Figure 9-10 shows that $d = 0.77$ for the 3-dB Chebyshev. With Eq. (9-11b) then, we find that

$$A_v = 3 - 0.77 = 2.23$$

Since the Op Amp is wired to work as a noninverting amplifier, its gain

$$A_v = \frac{R_f}{R_i} + 1$$

In this case,

$$2.23 = \frac{R_f}{R_i} + 1$$

or

$$\frac{R_F}{R_i} = 1.23$$

$R_f = 33$ kΩ and $R_i = 27$ kΩ would do nicely. Next, if we choose $C_1 = C_2 = C = 0.01$ μF, then, with Eq. (9-12b)

$$R = \frac{0.84}{2\pi (4 \text{ kHz})(0.01 \ \mu\text{F})} = 3342 \ \Omega; \text{ use } 3.3 \text{ k}\Omega$$

where the correction factor 0.84 was read from Table 9-1. If we can't find standard resistors that are close to calculated values of R, we can experiment with various available capacitor values until R works out more conveniently.

9.6 TYPICAL SECOND-ORDER HIGH PASS ACTIVE FILTER

The circuit of Fig. 9-12 is a second-order HP active filter. With proper selection of its components, its $A_{v(\text{dB})}$ vs. f curve has a 40 dB/decade slope at frequencies below the cutoff frequency f_c. As with LP filters, the component

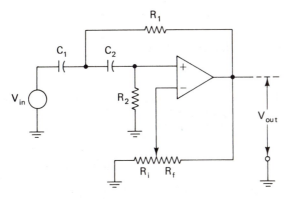

Figure 9-12 Second-order high pass active filter.

values determine the cutoff frequency and the damping coefficient which determines whether the filter is a Bessel, Butterworth, or Chebyshev type. In this case, the dampling coefficient

$$d = \frac{R_1C_1 + R_1C_2 + R_2C_2(1 - A_v)}{\sqrt{R_1R_2C_1C_2}} \tag{9-13}$$

and as with the LP filter, the HP filter's cutoff frequency

$$f_c = \frac{1}{2\,\pi\sqrt{R_1R_2C_1C_2}} \tag{9-8b}$$

If we take the equal-component approach—that is, let $R_1 = R_2 = R$ and $C_1 = C_2 = C$—the above equations simplify to

$$d = 3 - A_v \tag{9-11a}$$

which rearranged becomes

$$A_v = 3 - d \tag{9-11b}$$

and

$$f_c = \frac{1}{2\pi RC}$$

which, for Butterworth filters, can be shown to be

$$R = \frac{1}{2\,\pi\,f_cC} \tag{9-14a}$$

If designing Bessel or Chebyshev filters,

$$R = \frac{1}{(2\,\pi\,f_xC)K} \tag{9-14b}$$

where K is the correction factor listed in Table 9-1
      $f_x = f_c$ for the Bessel filter
or $f_x = f_p$ for the Chebyshev type

Example 9-7

We need second-order HP filters that significantly attenuate frequencies below 200 Hz. Design (a) a second-order Butterworth filter, letting the cutoff frequency $f_c = 200$ Hz; and (b) a second-order 1-dB Chebyshev type, letting the peak frequency $f_p = 200$ Hz.

Answers.   (a) Since $d = \sqrt{2} = 1.414$ for the Butterworth filter

$$A_v = 3 - 1.414 = 1.586$$

therefore

$$\frac{R_f}{R_i} = A_v - 1 = 0.586$$

If we select $R_f = 3.3$ kΩ, then $R_i = 5.6$ kΩ would suffice. Arbitrarily choosing $C_1 = C_2 = C = 0.2$ μF, then

$$R = \frac{1}{2 \pi (200 \text{ Hz})0.2 \text{ μF}} \cong 3.978 \text{ kΩ; use 3.9 kΩ} \qquad (9\text{-}14\text{a})$$

    (b) Since $d = 1$ for the 1-dB Chebyshev,

$$A_v = 3 - 1 = 2$$

and

$$\frac{R_f}{R_i} = A_v - 1 = 1$$

If we choose $C_1 = C_2 = C = 0.1$ μF, then

$$R = \frac{1}{[2 \pi (200 \text{ Hz})0.1 \text{ μF}]0.67} \cong 11.88 \text{ kΩ; use 12 kΩ} \quad (9\text{-}14\text{b})$$

Of course, the high frequencies that can pass through this filter are limited by the bandwidth of the Op Amp.*

9.7 ACTIVE BANDPASS FILTERS

A *bandpass filter* (BP) permits a range of frequencies to pass through it while it impedes all others. The circuit of Fig. 9-13a is called a multiple-feedback BP filter. With appropriate component values, its $A_{v(\text{dB})}$ vs. f response curve

---

   * See Appendix S for equations that determine phase shift caused by second-order active filters.

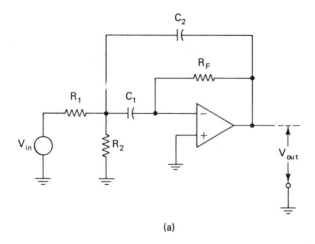

(a)

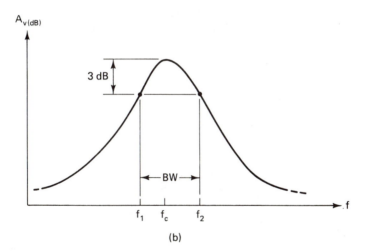

(b)

Figure 9-13 (a) Multiple feedback bandpass active filter, and (b) its gain vs. frequency response.

is as shown in part (b) of this figure. Note that the gain $A_{v(dB)}$ peaks at f_c—the *center frequency*. Also note that $A_{v(dB)}$ is 3 dB below its peak at frequencies f_1 and f_2. The frequencies between f_1 and f_2 are defined as being within the bandwidth (BW); that is,

$$BW = f_2 - f_1 \qquad (9\text{-}15)$$

The selectivity of a BP filter is usually described by its Q where

$$Q = \frac{f_c}{BW} \qquad (9\text{-}16a)$$

or

$$BW = \frac{f_c}{Q} \qquad (9\text{-}16b)$$

This last equation shows that larger values of Q yield smaller (narrower) bandwidths. For this circuit of Fig. 9-13, obtainable values of Q lie between 1 and 20. Also, at the center frequency f_c, the gain must be less than twice the Q squared; that is

$$A_{v_c} < 2Q^2 \qquad (9\text{-}17)$$

where

$$A_{v_c} = \frac{V_{out}}{V_{in}} \text{ at } f_c$$

As with the LP and HP filters, filter design is simplified if we make some of the filter components equal. In this case, if $C_1 = C_2 = C$, then

$$R_1 = \frac{Q}{2 \pi f_c A_{v_c} C} \qquad (9\text{-}18)$$

$$R_2 = \frac{Q}{2 \pi f_c C(2Q^2 - A_{v_c})} \qquad (9\text{-}19)$$

and

$$R_F = \frac{2Q}{2 \pi f_c C} \qquad (9\text{-}20)$$

Example 9-8

Design a BP filter that has a center frequency of a 1 kHz and a $Q = 10$. Make the center frequency gain $A_{v_c} = 2$.

Answers.   Arbitrarily choosing $C = 0.02 \ \mu F$,

$$R_1 = \frac{10}{2 \pi (1 \text{ kHz})2(0.02 \ \mu F)} = 39.8 \text{ k}\Omega; \text{ use } 40 \text{ k}\Omega \qquad (9\text{-}18)$$

$$R_2 = \frac{10}{2 \pi (1 \text{ kHz})0.02 \ \mu F[2(10)^2 - 2]} = 401.9 \ \Omega; \text{ use } 400 \ \Omega \qquad (9\text{-}19)$$

and

$$R_F = \frac{2(10)}{2 \pi (1 \text{ kHz})0.02 \ \mu F} = 159 \text{ k}\Omega; \text{ use } 160 \text{ k}\Omega \qquad (9\text{-}20)$$

In terms of the multiple-feedback filter's component values, its center frequency

$$f_c = \frac{1}{2\pi} \sqrt{\frac{R_1 + R_2}{R_1 R_2 R_F C^2}} \qquad (9\text{-}21)$$

Generally, this circuit's gain A_v can be found with the equation

$$A_v = \frac{(A_{v_c} d\omega_c \omega)^2}{\omega^4 + \omega_c^2(d^2 - 2)\omega^2 + \omega_c^4} \qquad (9\text{-}22)$$

where

$$\omega = 2\pi f$$

$$\omega_c = 2\pi f_c$$

$$d = 1/Q$$

and

$$A_{v_c} = \frac{R_F}{2R_1} \text{ if } C_1 = C_2 \qquad (9\text{-}23)$$

The answers to the previous example can be verified with these equations.*

Since the bandwidth is determined by the Q, the 3-dB-down frequencies, which mark the bounds of the BW, can be determined with the following equations:

$$f_1 = f_c \left[\sqrt{\frac{1}{4Q^2} + 1} - \frac{1}{2Q} \right] \qquad (9\text{-}24\text{a})$$

and

$$f_2 = f_c \left[\sqrt{\frac{1}{4Q^2} + 1} + \frac{1}{2Q} \right] \qquad (9\text{-}24\text{b})$$

9.8 ACTIVE BANDSTOP (NOTCH) FILTERS

Often, in audio and medical equipment, ac power frequencies are a source of unwanted noise. In such cases, bandstop filters, also called *band-reject* or *notch filters*, can serve to remove the 50-Hz, 60-Hz or 400-Hz noise while allowing most of the signal frequencies to pass through.

The circuit shown in Fig. 9-14a is a multiple-feedback notch filter. The first stage is a bandpass filter like the one of Fig. 9-13. The second stage is a summing circuit. Signals V_{in} and V_{out} are applied to the two inputs of this

* See Appendix P for a BASIC program that prints a table of gain vs. frequency characteristics for the multiple-feedback bandpass active filter.

summer. These signals are out of phase and tend to cancel each other's effects via the summing amplifier. Maximum cancellation occurs at the center frequency f_c resulting in a $A_{v(dB)}$ vs. f curve as shown Fig. 9-14b.

The components R_1, R_2 and R_F in the notch filter are selected as for the bandpass filter; that is, use Eqs. (9-18), (9-19), and (9-20). The ratio R_5/R_4 determines the gain "seen" by signal V_{out} "looking" into the summing amplifier. Similarly, the summer amplifies the signal V_{in} by the ratio R_5/R_3.

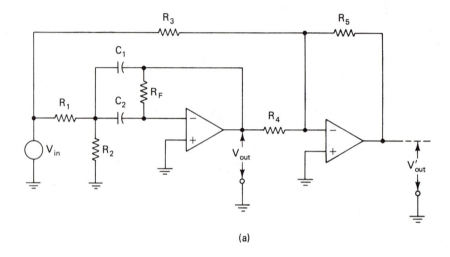

(a)

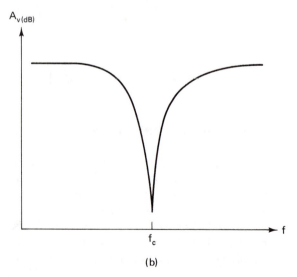

(b)

Figure 9-14 (a) Multiple feedback bandstop (notch) filter, and (b) its gain vs. frequency response.

Maximum cancellation, and greatest depth of the notch, occur at f_c if

$$A_{v_c}\left(\frac{R_5}{R_4}\right) = \frac{R_5}{R_3} \tag{9-25a}$$

where

$$A_{v_c} = \frac{R_F}{2R_1} \tag{9-23}$$

Equation 9-25a rearranged becomes

$$R_3 = \frac{R_4}{A_{v_c}} \tag{9-25b}$$

Resistance R_3 can be a potentiometer for control of the depth of the notch.

Example 9-9

Design a notch filter with a center frequency $f_c = 60$ Hz. Use $Q = 10$ and $A_{v_c} = 2$ in the design Eqs. (9-18) through (9-20).

Answers.   Choosing an available capacitor, like $C = 0.1$ μF, we can show that in the first stage

$$R_1 = \frac{10}{2\,\pi(60)2(0.1\ \mu\text{F})} = 132.63\text{ k}\Omega;\text{ use } 130\text{ k}\Omega \tag{9-18}$$

$$R_2 = \frac{10}{2\,\pi(60)0.1\ \mu\text{F}[2(10^2) - 2]} = 1.3397\text{ k}\Omega;\text{ use } 1.3\text{ k}\Omega \tag{9-19}$$

and

$$R_F = \frac{2(10)}{2\,\pi(60)0.1\ \mu\text{F}} = 530.52\text{ k}\Omega;\text{ use } 560\text{ k}\Omega \tag{9-20}$$

Arbitrarily choosing a ratio $R_5/R_4 = 10$, let $R_5 = 10$ kΩ and $R_4 = 1$ kΩ. Then, by Eq. (9-25b),

$$R_3 = 1\text{ k}\Omega/2 = 500\ \Omega;\text{ use } 470\ \Omega^* \tag{9-25b}$$

REVIEW QUESTIONS

9-1. What is meant by the term *cutoff frequency* as it is used with high pass and low pass filters?

9-2. How many resistors and capacitors does a first-order active filter have?

* See Appendix Q for a BASIC program that prints a table of gain vs. frequency characteristics for the multiple feedback notch filter.

9-3. How do the values of resistance and capacitive reactance compare at the cutoff frequency in a first-order active filter?

9-4. What is the slope of a first-order LP active filter's $A_{v(dB)}$ vs. f curve at frequencies well above the cutoff frequency?

9-5. What is the slope of a first-order HP active filter's $A_{v(dB)}$ vs. f curve at frequencies well below the cutoff frequency?

9-6. What is the slope of a second-order LP filter's $A_{v(dB)}$ vs. f curve at frequencies significantly higher than the cutoff frequency?

9-7. What is the slope of a third-order HP filter's $A_{v(dB)}$ vs. f curve at frequencies significantly below the cutoff frequency?

9-8. How many resistors and capacitors are needed in a second-order LP active filter? How many are needed in a second-order HP active filter?

9-9. A filter section of an audio equalizer consists of two cascaded second-order LP active filters; each has the same cutoff frequency f_c. At what rate does the $A_{v(dB)}$ vs. f curve roll-off in the range of ten times f_c?

9-10. What parameter distinguishes the Chebyshev filters from Butterworth types?

9-11. What kind of filter is the circuit of Fig 9-15b? Specify if it is a first-order, second-order, etc., and if it is a low pass, high pass, bandpass, etc.

9-12. Comparing the circuits (a) and (b) of Fig. 9-15, which is more suited for the equal component design? Why?

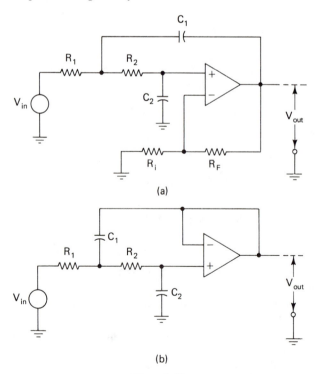

(a)

(b)

Figure 9-15

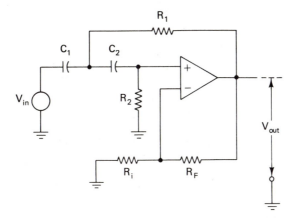

Figure 9-16

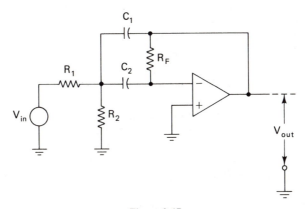

Figure 9-17

9-13. What kind of filter is the circuit of Fig. 9-16?

9-14. What kind of filter is the circuit of Fig. 9-17?

9-15. Sketch modified versions of the circuits in Fig. 9-3 so that each has a gain adjustment.

9-16. How would you modify the circuit of Fig. 9-5 to minimize the drift in the quiescent value of V_{out} caused by drifting bias currents?

9-17. What is meant by the term *bandwidth* in reference to a bandpass filter?

9-18. What is the center frequency in a bandpass filter? In a bandstop filter?

PROBLEMS

9-1. Referring to the circuit of Fig 9-3b, if $R = 1.2$ kΩ and $C = 0.22$ μF, what is its cutoff frequency? To minimize drift of V_{out}, due to drifting input bias current, what value of R_F should we use?

9-2. Referring to the circuit of Fig. 9-3a, if $R = 1.8$ kΩ and $C = 0.025$ μF in the circuit of Fig. 9-3, what is the cutoff frequency? What value of R_F will minimize the drift of V_{out} due to drifting input bias current?

9-3. Referring to the circuit described in Prob. 9-1, what is the gain, in dB, at frequencies (a) 200 Hz, (b) 400 Hz, (c) 600 Hz, (d) 800 Hz, and (e) 1 kHz?

9-4. What is the gain, in dB, of the circuit described in Prob. 9-2 at each of the following frequencies? (a) 2.5 kHz, (b) 3 kHz, (c) 3.5 kHz, (d) 4 kHz and (e) 4.5 kHz.

9-5. Sketch an approximate $A_{v(dB)}$ vs. f curve of the circuit described in Prob. 9-1. Include the cutoff frequency and assume that frequency is plotted on a logrithmic scale.

9-6. Sketch an approximate $A_{v(dB)}$ vs. f curve of the circuit described in Prob. 9-2. Include the cutoff frequency and assume that frequency is plotted on a logrithmic scale.

9-7. If, in the circuit of Fig. 9-15a, $R_1 = R_2 = 1.8$ kΩ, $R_i = 1$ kΩ, $R_F = 1.1$ kΩ and $C_1 = C_2 = 0.047$ μF, what is its dampling coefficient? What kind of filter is this?

9-8. If $R_1 = R_2 = 4$ kΩ, $R_i = 5.6$ kΩ, $R_F = 3.3$ kΩ and $C_1 = C_2 = 0.02$ μF in the circuit of Fig. 9-15a, what is its damping coefficient? What kind of filter is it?

9-9. Referring to the circuit described in Prob. 9-7, if you found it to be a Bessel or Butterworth filter, find its cutoff frequency f_c. If you determined that this filter is a Chebyshev type, find its peak frequency f_p. *Hint:* Rearrange Eq. (9-12b) if solving for f_p.

9-10. Referring to the circuit described in Prob. 9-8, determine its cutoff frequency f_c if it is a Bessel or Butterworth type, or find its peak frequency f_p if it is a Chebyshev type. (*Hint:* Rearrange Eq. (9-12b) if solving for f_p.)

9-11. Using the circuit type found in Fig. 9-15b, select components so that it works as a Butterworth LP filter that has a cutoff frequency of 800 Hz. Use $C_1 = 0.2$ μF.

9-12. Using the circuit type (a) of Fig. 9-15, select components so that it works as a 1-dB Chebyshev LP filter that has a peak frequency of 1 kHz. Use $C_1 = 0.02$ μF.

9-13. If $R_1 = R_2 = 16$ kΩ, $R_i = 22$ kΩ, $R_F = 27$ kΩ, and $C_1 = C_2 = 0.047$ μF, in the circuit of Fig. 9-16, what is its dampling coefficient? What kind of filter is it?

9-14. If in the circuit of Fig. 9-16, $R_1 = R_2 = 20$ kΩ, $R_i = 39$ kΩ, $R_F = 11$ kΩ and $C_1 = C_2 = 0.1$ μF, what is the dampling coefficient? What kind of filter is it?

9-15. Referring to the circuit described in Prob. 9-13, determine its cutoff frequency f_c if it is a Bessel or Butterworth type, or find its peak frequency f_p if it is a Chebyshev type. *Hint:* Rearrange Eq. (9-14b) if solving for f_p.

9-16. Referring to the circuit described in Prob. 9-14, determine its cutoff frequency f_c if it is a Bessel or Butterworth type, or find its peak frequency f_p if it is a Chebyshev type. *Hint:* Rearrange Eq. (9-14b) if solving for f_p.

9-17. Select components for the circuit of Fig. 9-17 so that it works as a bandpass filter with a center frequency f_c of 1 kHz, a gain of 10 at f_c and a Q of 10. Use $C_1 = C_2 = C = 0.02$ μF. If your design meets these specifications, what is its bandwidth and at what frequencies does the gain $A_{v(dB)}$ drop 3 dB below its value at f_c?

9-18. Select components for the circuit of Fig. 9-17 so that it works as a bandpass filter with a center frequency f_c of 200 Hz, a gain of 2 at f_c and a Q of 8. Use C_1 = C_2 = C = 0.01 μF. If your design meets these specifications, what is its bandwidth and at what frequencies does its gain $A_{v(dB)}$ drop 3 dB below the gain at f_c?

9-19. Referring to the circuit of Fig. 9-17, if R_1 = 8.2 kΩ, R_2 = 390 Ω, R_F = 160 kΩ and C_1 = C_2 = 0.02 μF, what are its center frequency f_c and gain at the center frequency?

9-20. If, in the circuit of Fig. 9-17, R_1 = 1.3 kΩ, R_2 = 33 Ω, R_F = 13 kΩ and C_1 = C_2 = 0.1 μF, what are its center frequency f_c and gain in decibels at f_c?

9-21. Select components for the circuit of Fig. 9-14 so that it works as a notch filter with a center frequency of 50 Hz. Use Q = 5 and A_{v_c} = 1 in the design Eqs. (9-18) through (9-20). Let C_1 = C_2 = 0.4 μF and the ratio R_5/R_4 = 10.

9-22. Select components for the circuit of Fig. 9-14 so that it works as a notch filter with a center frequency of 400 Hz. Use Q = 8 and A_{v_c} = 2 in the design Eqs. (9-18) through (9-20). Let C_1 = C_2 = 0.1 μF and the ratio R_5/R_4 = 10.

10

NONLINEAR AND DIGITAL
APPLICATIONS OF OP AMPS

In some applications, Op Amps are not required to work in a linear mode. Instead, they are required to switch abruptly from one output voltage level to another even though the input voltage changes are gradual. Occasionally the output voltage swing of an Op Amp is to be kept within specific limits. In such cases, Op Amps are used with externally wired components that clip output signals which attempt to swing beyond the predetermined limits. Beyond these, there are limitless possible nonlinear applications of Op Amps. The analyses of a few in this chapter will give us insight into other possibilities.

10.1 VOLTAGE LIMITERS

The output swing of a general-purpose Op Amp is often too large for the inputs of some circuits. Digital circuits, for example, usually require specific levels of inputs that are not, without modification, available from the output of a typical Op Amp. Several output-voltage-limiting circuits are shown in Fig. 10-1 through 10-5.

In the circuit of Fig. 10-1, the zener diodes D_1 and D_2 limit the peak-to-peak value of the output voltage V_o. The zener voltage V_{z_1} of D_1 is approximately equal to the maximum possible positive value of V_o—i.e., we cannot drive the output V_o more positive than about the V_{z_1} voltage. The zener voltage V_{z_2} of D_2 similarly is about equal to the maximum negative value of V_o.

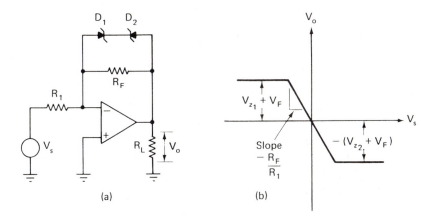

Figure 10-1 (a) Positive and negative voltage clipper and (b) its transfer characteristics.

More specifically, as shown by the transfer characteristic in Fig. 10-1b, the output V_o levels off (is clipped) at a voltage $V_{z_1} + V_F$ as it swings positively. On the other hand, as the output V_o swings negatively, it is clipped at $-(V_{z_2} + V_F)$. The voltage V_F is the voltage drop across the *forward*-biased zener and is typically about 0.7 V. Of course, when V_o swings positively, D_1 is reverse-biased and D_2 is forward-biased; when V_o swings negatively, D_1 and D_2 are forward- and reverse-biased, respectively.

Example 10-1

In the circuit of Fig. 10-1, suppose that each zener has a 6.3-V zener voltage and a 0.7-V forward drop, and that $R_1 = 1$ kΩ and $R_F = 20$ kΩ. Describe the output waveforms with each of the following sine-wave inputs:

(a) $V_s = 0.3$ V peak,
(b) $V_s = 0.6$ V peak, and
(c) $V_s = 3$ V peak.

Answer.   The gain A_v of this stage is about $-R_F/R_1 = -20$, and its output peaks are limited to

$$V_{z_1} + V_F = 6.3 + 0.7 = 7 \text{ V}$$

and on positive alternations to

$$-(V_{z_2} + V_F) = -(6.3 + 0.7) = -7 \text{ V}$$

on the negative alternations.

(a) When $V_s = 0.3$ V peak, $V_o = A_v V_s = -20(0.3) = -6$ V peak. This output V_o does not force either zener into zener (reverse) conduction, and therefore the output waveform is unclipped and sinusoidal.

(b) When $V_s = 0.6$ V peak, the output would peak to $-20(0.6) = -12$ V if the zeners were not across R_F. Since they are, V_o gets clipped at 7 V on positive and negative alternations.

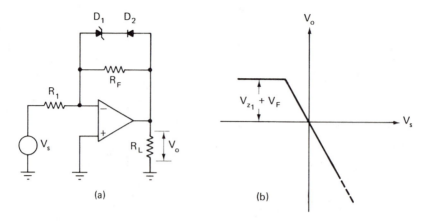

Figure 10-2   (a) Positive voltage clipper and (b) its transfer characteristics.

(c) When $V_s = 3$ V peak, the output would peak to $-20(3) = -60$ V if limiting factors were not present. Even without the zeners, the general-purpose Op Amp will saturate if we attempt to drive it this hard. With the zeners, the output is clipped at about 7 V on each alternation, and the output V_o has the appearance of a squarewave.

The zener and rectifier diodes in the circuit of Fig. 10-2 limit the swing of the output voltage V_o in a positive direction only, as shown by its transfer characteristic in Fig. 10-2b. This means that, if V_o attempts to rise above the voltage $V_{z_1} + V_F$, the zener D_1 goes into zener (avalanche) conduction and the waveform of V_o is clipped. The negative alternations are not clipped unless the Op Amp is driven into negative saturation. The circuit in Fig. 10-3 is similar except that its zener and rectifier diodes limit the output V_o to $-(V_{z_2} + V_F)$ when it swings in the negative direction. V_o can swing positively to the point where the Op Amp positively saturates.

If only a single zener is used with no rectifier, as in Figs. 10-4 and 10-5,

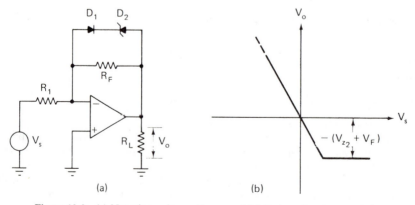

Figure 10-3   (a) Negative voltage clipper and (b) its transfer characteristics.

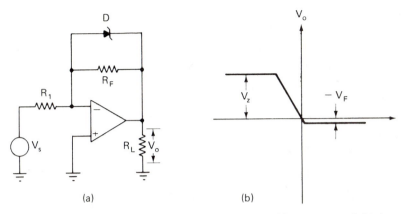

Figure 10-4 (a) Half-wave rectifier with limited positive output and (b) its transfer characteristics.

and a sinewave input V_s is applied, the swing in the output V_o is limited by the zener voltage on half of the alternations and by the zener's forward drop V_F on the remaining alternations.

Example 10-2

Referring to the circuits in Figs. 10-1 through 10-5, suppose that in each, $R_1 = 1\ k\Omega$, $R_F = 10\ k\Omega$, $V_z = 4\ V$ of each zener, and that the drop across each forward-biased diode and zener is negligible. Sketch the output V_o of each circuit if a 2-V peak, 60-Hz sinewave input voltage V_s is applied to its input.

Answer.   See Fig. 10-6a. This waveform is the input V_s applied to each circuit. Each circuit's gain $A_v = -R_F/R_1 = -10$. Thus the outputs V_o are out of phase with V_s and they *tend* to peak at 20 V on positive and negative alternations. Clipping occurs however. The waveform shown in Fig. 10-6b is the output V_o of the circuit in Fig. 10-1. The waveform in c of Fig. 10-6 is the output V_o of the circuit in Fig. 10-2. The

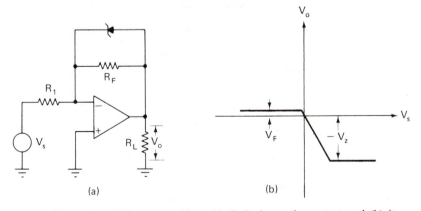

Figure 10-5 (a) Half-wave rectifier with limited negative output and (b) its transfer characteristics.

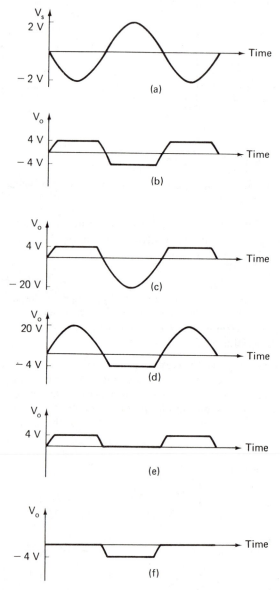

Figure 10-6 If (a) is the input waveform V_s, $R_F/R_1 = 10$, and $V_z = 4\ V$, then (b) is the output V_o of the circuit in Fig. 9-1, (c) is the output V_o of the circuit in Fig. 9-2, (d) is the output V_o of the circuit in Fig. 9-3, (e) is the output V_o of the circuit in Fig. 9-4, and (f) is the output V_o of the circuit in Fig. 9-5.

waveform in Fig. 10-6d is the output V_o of the circuit in Fig. 10-3. In Fig. 10-6e, the waveform is the output V_o of the circuit in Fig. 10-4. And the waveform in Fig. 10-6f is the output V_o of the circuit in Fig. 10-5.

10.2 THE ZERO-CROSSING DETECTOR

When used in open loop, the Op Amp is very sensitive to changes in input voltages. In fact, fractions of millivolts can easily drive the Op Amp into saturation when no feedback is used. This feature is an advantage in some applications. For example, in Fig. 10-7a, the circuit works as a zero-crossing detector or a sine-wave to squarewave converter. As shown in its output V_o vs. input V_s waveforms in Fig. 10-7b, the output V_o swings and saturates negatively when the input V_s passes through zero in the positive direction. On the other hand, when V_s passes through zero negatively, the Op Amp's output V_o is driven into positive saturation.

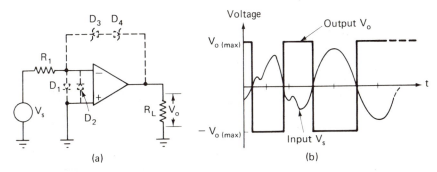

Figure 10-7 (a) The zero-crossing detector and (b) its typical waveforms. D_1 and D_2 are input-protecting diodes and D_3 and D_4 are output-limiting zeners.

If the peak of the input voltage V_s exceeds a volt or so, the input diodes D_1 and D_2 protect the Op Amp. Of course, if the Op Amp is a type that is input-protected, which means that the equivalent of diodes D_1 and D_2 are built internally, the input diodes are usually unnecessary. If the output swing between $V_{o(max)}$ and $-V_{o(max)}$ is excessive, as it would be for many digital circuit inputs, either or both of the zeners D_3 or D_4 can be used to clip V_o at whatever limits are necessary. The resistance R_1 limits the current through the input protection diodes. A large R_1, however, adds to offset errors. A resistance equal to R_1 can be placed between the noninverting input and ground. Such a resistor reduces offset problems and is more necessary when large values of R_1 are used and if the input bias current is relatively large as it is with general-purpose Op Amps.

10.3 OP AMPS AS COMPARATORS

The comparator, as its name implies, compares two voltages. One is usually a fixed reference voltage, V_R, and the other a time-varying signal voltage which is often called an *analog* voltage V_A. The Op Amp comparator circuit is very similar to the zero-crossing detector discussed in the previous section, except that a reference voltage V_R is used between one input and ground on the comparator (see Fig. 10-8). This causes the Op Amp's output to swing from $V_{o(max)}$ to $-V_{o(max)}$, or vice versa, as the analog voltage V_A passes through the reference voltage value V_R. This reference V_R can be either positive or negative with respect to ground, and of course, its value and polarity determine the V_A voltage that causes the output to switch. Note in Fig. 10-8b that, with the analog input waveform, the output has the waveform in Fig. 10-8c or d, depending on whether V_R is positive or negative, respectively. Obviously, the amplitude of V_A must be large enough to pass through V_R if the switching action is to take place. The circuit in Fig. 10-8 is an inverting type. A noninverting Op Amp comparator and its typical

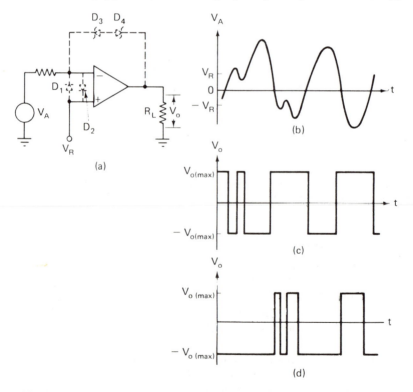

Figure 10-8 (a) An Op Amp as a comparator. With input waveform (b), the output can be waveform (c) with a positive reference voltage, or waveform (d) if the reference is negative.

waveforms are shown in Fig. 10-9. A resistor $R_2 = 200 \, \Omega$ or so and one or both zeners D_3 or D_4 are used if necessary to keep the output swing within required limits.

An Op Amp comparator that has hysteresis in its transfer function is shown in Fig. 10-10. In b of this figure, the transfer function shows that the output voltage swings from $V_{o(max)}$ to $-V_{o(max)}$ when the input voltage V_A crosses through V_a while increasing positively. However, note that the output swings from $-V_{o(max)}$ to $V_{o(max)}$ when the input V_A crosses through V_a' when increasing negatively. This comparator, therefore, has two reference voltages; one positive (V_a) and one negative (V_a'), whereas the circuit

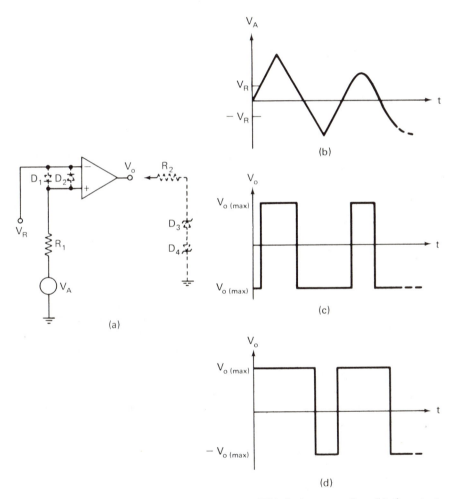

Figure 10-9   (a) Noninverting Op Amp comparator. With the input waveform (b), the output can be waveform (c) if the reference voltage is positive or waveform (d) if the reference is negative.

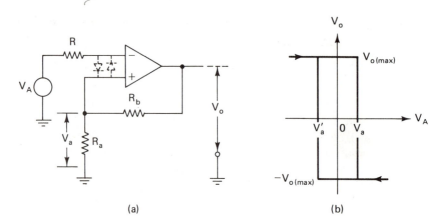

Figure 10-10 (a) Op Amp comparator with hysteresis and (b) its transfer function.

of Fig. 10-8 has only one reference voltage V_R. The switching of the output V_o at different values of V_A; depending on whether V_A is changing in the positive or the negative direction, is called *hysteresis* and the circuit itself is called a *schmitt trigger*.

When the output is at $V_{o(max)}$, the voltage at the noninverting input 2 is positive, having a value

$$V_a = \frac{V_{o(max)}R_a}{R_a + R_b} \tag{10-1a}$$

On the other hand, when the output is at $-V_{o(max)}$, the voltage at input 2 is negative and has the value

$$V_a' = \frac{-V_{o(max)}R_a}{R_a + R_b} \tag{10-1b}$$

The difference in the voltages V_a and V_a' is sometimes called the *noise margin*. When noise is expected as part of the analog input voltage V_A, hysteresis can effectively eliminate undesirable switching of the output V_o due to such noise.

Example 10-3

(a) If in the schmitt trigger circuit of Fig. 10-10, $R_a = 1$ kΩ, $R_b = 9$ kΩ, the Op Amp saturates at ± 10 V, and the input signal V_A has the waveform of Fig. 10-11, sketch the resulting output.

(b) If this input signal V_A is applied to the circuit of Fig. 10-8 while its reference $V_A = 1$ V, sketch the resulting output.

Answers.   (a) With the component values specified,

$$V_a = \frac{10\ \text{V}(1\ \text{k}\Omega)}{1\ \text{k}\Omega + 9\ \text{k}\Omega} = 1\ \text{V} \tag{10-1a}$$

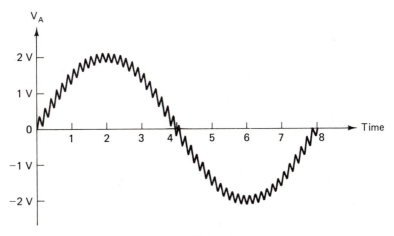

Figure 10-11 Noisy sinewave.

and

$$V_a' = \frac{-10 \text{ V}(1 \text{ k}\Omega)}{1 \text{ k}\Omega + 9 \text{ k}\Omega} = -1 \text{ V}$$

As shown in Fig. 10-12a, V_o switches to -10 V when V_A crosses through 1 V when going in the positive direction. This output V_o then switches to 10 V when V_A crosses through -1 V while changing in the negative direction. The noise margin is 2 V.

(b) Note in Fig. 10-12b that V_o switches *each* time V_A crosses through 1 V and that the noise affects the waveform of V_o.

The threshold voltages at which the schmitt trigger circuit switches can be shifted by use of a reference voltage V_{ref}, instead of ground, at one end of R_2 (see Fig. 10-13). As indicated by its transfer characteristic shown in Fig. 10-13b, the output is positively saturated as long as the input signal voltage V_s is less than the *upper threshold voltage* V_a. If V_A rises slightly above this threshold voltage V_a, the output swings to $-V_{o(\max)}$ and stays there until V_A drops below a *lower threshold voltage* V_a'. The threshold voltages are determined by the components R_1, R_2, and the dc reference voltage V_{ref}. These threshold voltages can be determined with the equation

$$V_a = \frac{R_2(V_o - V_{\text{ref}})}{R_1 + R_2} + V_{\text{ref}} \qquad (10\text{-}2)$$

where V_o is the maximum positive output voltage when solving for the
    upper threshold, or
    V_o is the maximum negative output voltage when solving for the
    lower threshold.

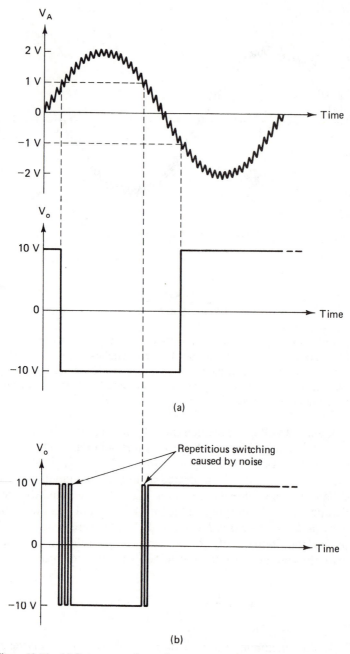

Figure 10-12 (a) Output waveform when comparator has hysteresis in its transfer function, (b) output waveform has repetitious switching caused by noise if comparator has no hysteresis.

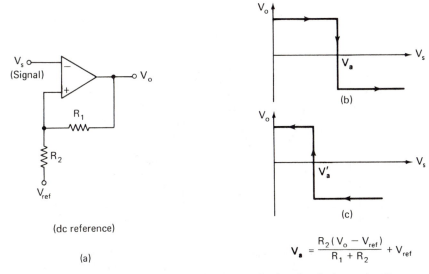

(dc reference)

(a)

$$V_a = \frac{R_2(V_o - V_{ref})}{R_1 + R_2} + V_{ref}$$

Figure 10-13   (a) Schmitt trigger circuit, (b) transfer function for increasing V_s, and (c) transfer function for decreasing V_s.

Example 10-4

If in the circuit of Fig. 10-13, $R_1 = 10$ kΩ, $R_2 = 220$ Ω, $V_{ref} = 2$ V, and the Op Amp saturates at $V_o = \pm 10$ V, what are the upper and lower threshold voltages?

Answer.   Since the Op Amp saturates positively at $+10$ V, the upper threshold voltage is

$$V_a = \frac{220\ \Omega(10\ \text{V} - 2\ \text{V})}{10.22\ \text{k}\Omega} + 2\ \text{V} \cong 2.17\ \text{V}$$

This means that if the input voltage V_s rises slightly higher than 2.17 V, the Op Amp is driven into negative saturation, -10 V in this case.

   With the output V_o as negative as -10 V, the lower threshold voltage is

$$V_a' = \frac{220\ \Omega(-10\ \text{V} - 2\ \text{V})}{10.22\ \text{k}\Omega} + 2\ \text{V} \cong 1.74\ \text{V}$$

This means that if the input V_s drops slightly below 1.74 V, the output swings back to $+10$ V. Output vs. input waveforms of this circuit are shown in Fig. 10-14.

10.4 DIGITAL-TO-ANALOG (D/A) CONVERTERS

Digital systems usually work with two levels of voltage referred to as HIGH and LOW signals or as logic 1 and logic 0. This two-level approach to performing computations and decisions is called a *binary system*, and values

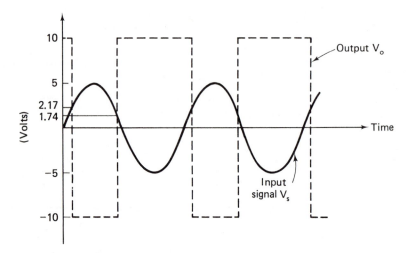

Figure 10-14 Output V_o vs. input V_s waveforms of the Schmitt trigger if the upper threshold voltage $V_a = 2.17$ V and the lower threshold $V_a' = 1.74$ V.

in such a system are expressed and processed in binary numbers. Some binary numbers along with their decimal equivalents are shown in Table 10-1. Such binary numbers are often read off of flip-flops followed by level amplifiers which can be viewed here as being equivalent to switches that are capable of providing *either* an output voltage V or 0 V as shown in Fig. 10-15.

Decimal	Binary
	DCBA
0	0000
1	0001
2	0010
3	0011
4	0100
5	0101
6	0110
7	0111
8	1000
9	1001
10	1010
11	1011
12	1100
13	1101
14	1110
15	1111
16	10000

Table 10-1

Binary output

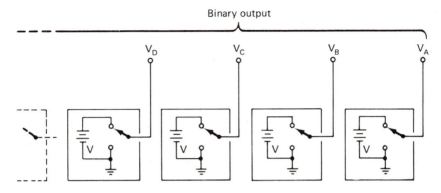

Figure 10-15   Equivalent of a digital system.

When it is necessary to convert a binary output from a digital system to some equivalent analog voltage, a digital-to-analog (D/A) converter is used. An analog output should have 16 possible values if there are 16 combinations of digital voltages V_A through V_D. For example, since the binary number 0110 (decimal 6) is twice the value of the binary number 0011 (decimal 3), an analog equivalent voltage of 0110 is double the analog voltage representing 0011.

Binary outputs from digital systems can be converted to analog equivalent voltages by the use of either of the resistive networks of Fig. 10-16. The binary-weighted network in Fig. 10-16a, though requiring fewer

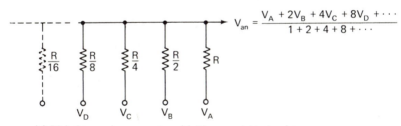

$$V_{an} = \frac{V_A + 2V_B + 4V_C + 8V_D + \cdots}{1 + 2 + 4 + 8 + \cdots}$$

(a) Digital-to-analog converter with binary-weighted resistors.

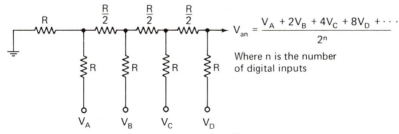

$$V_{an} = \frac{V_A + 2V_B + 4V_C + 8V_D + \cdots}{2^n}$$

Where n is the number of digital inputs

(b) Digital-to-analog converter with R and $\frac{R}{2}$ resistors.

Figure 10-16   Resistive networks as digital-to-analog (D/A) converters.

resistors, needs a variety of precision resistance values. Its analog output V_{an}, as a function of the two-level inputs, can be determined with the equation:

$$V_{an} = \frac{V_A + 2V_B + 4V_C + 8V_D + \cdots}{1 + 2 + 4 + 8 + \cdots} \tag{10-3}$$

If each of its digital inputs, V_A through V_D, is either 15 V or 0 V, the analog outputs V_{an} versus all possible combinations of inputs are shown in Table 10-2.

The D/A converter of Fig. 10-16b, using R and $R/2$ resistors, requires more resistors but only two sets of precision resistance values. Its output is

$$V_{an} = \frac{V_A + 2V_B + 4V_C + 8V_D + \cdots}{2^n} \tag{10-4}$$

Decimal equivalent

	V_D	V_C	V_B	V_A	V_{an}
0	0 V	0 V	0 V	0 V	0 V
1	0 V	0 V	0 V	15 V	1 V
2	0 V	0 V	15 V	0 V	2 V
3	0 V	0 V	15 V	15 V	3 V
4	0 V	15 V	0 V	0 V	4 V
5	0 V	15 V	0 V	15 V	5 V
6	0 V	15 V	15 V	0 V	6 V
7	0 V	15 V	15 V	15 V	7 V
8	15 V	0 V	0 V	0 V	8 V
9	15 V	0 V	0 V	15 V	9 V
10	15 V	0 V	15 V	0 V	10 V
11	15 V	0 V	15 V	15 V	11 V
12	15 V	15 V	0 V	0 V	12 V
13	15 V	15 V	0 V	15 V	13 V
14	15 V	15 V	15 V	0 V	14 V
15	15 V	15 V	15 V	15 V	15 V

Decimal equivalent

	V_D	V_C	V_B	V_A	V_{an}
0	0 V	0 V	0 V	0 V	0 V
1	0 V	0 V	0 V	8 V	0.5 V
2	0 V	0 V	8 V	0 V	1 V
3	0 V	0 V	8 V	8 V	1.5 V
4	0 V	8 V	0 V	0 V	2 V
5	0 V	8 V	0 V	8 V	2.5 V
6	0 V	8 V	8 V	0 V	3 V
7	0 V	8 V	8 V	8 V	3.5 V
8	8 V	0 V	0 V	0 V	4 V
9	8 V	0 V	0 V	8 V	4.5 V
10	8 V	0 V	8 V	0 V	5 V
11	8 V	0 V	8 V	8 V	5.5 V
12	8 V	8 V	0 V	0 V	6 V
13	8 V	8 V	0 V	8 V	6.5 V
14	8 V	8 V	8 V	0 V	7 V
15	8 V	8 V	8 V	8 V	7.5 V

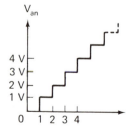

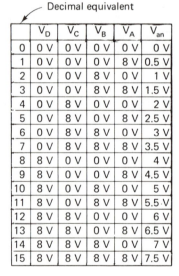

Table 10-2 Output vs inputs of the binary-weighted, resistive network where the digital logic levels are 0 V and 15 V.

Table 10-3 Output vs inputs of the R and $R/2$ resistive network where the digital logic levels are 0 V and 8 V.

where n is the number of digital inputs. If the levels of each of the inputs are 8 V and 0 V, all possible analog outputs of this four-input resistive network are shown in Table 10-3.

If the digital inputs to either network increase consecutively through larger decimal equivalents, the analog outputs are staircase waveforms as shown in Table 10-2 and 10-3.

Both D/A converters of Fig. 10-16 must work into large-resistance loads; otherwise, the accuracy of their equations, (10-3) and (10-4), degenerates. Therefore, as shown in Fig. 10-17, an Op Amp is well suited to buffer a D/A resistive network to a low-resistance load and to provide gain as well.

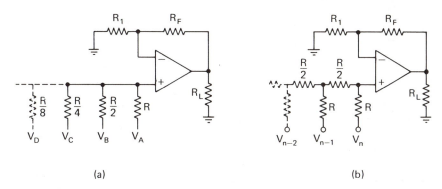

(a)   (b)

Figure 10-17   Digital-to-analog (D/A) converters with buffered output and gain.

10.5 ANALOG-TO-DIGITAL (A/D) CONVERTERS

Often an analog voltage must be converted to a digital equivalent, such as in a digital voltmeter. In such cases, the principle of the previously discussed digital-to-analog D/A converter can be reversed to perform analog-to-digital A/D conversion. A system that converts analog voltages into digital equivalents is shown in Fig. 10-18.

The up-down counter in the system of Fig. 10-18 has a digital output that increases with each clock pulse when its "count-up" line is HIGH and its "count-down" line is LOW. On the other hand, its digital output decreases with each clock pulse when its count-up line is LOW and its count-down line is HIGH.

The Op Amp U_2 works as a comparator. When its output is HIGH (positively saturated), the count-up line is also HIGH. Therefore, when U_2's output is LOW (negatively saturated), the count-up line is LOW too. Thus, depending on whether U_2's output is HIGH or LOW, the up-down counter counts digitally up or down, respectively. When the up-down counter is counting up, an upward staircase voltage appears at point a. When the counter is counting downward, a downward staircase is present at point a.

Since the Op Amp U_2 is operated in open loop, its output goes HIGH

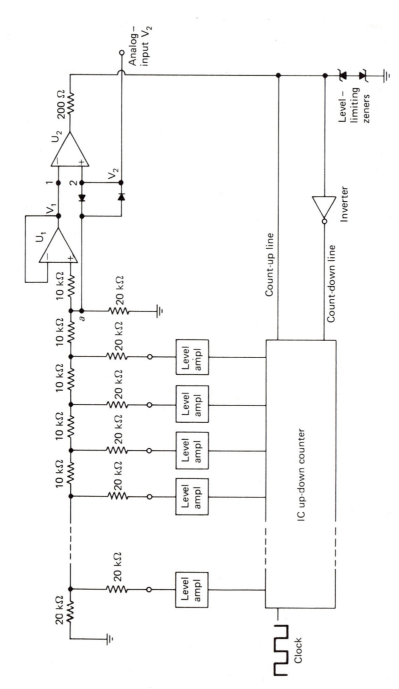

Figure 10-18 Analog-to-digital (A/D) converter.

when its input 1 becomes *slightly* more negative than input 2. Conversely, its output goes LOW when its input 1 becomes *slightly* more positive than input 2.

The first Op Amp U_1 is wired to work as a voltage follower and it buffers the D/A resistive network. Essentially then, the output voltage of the resistive network is applied to input 1 of U_2 and is thus compared with the analog input voltage V_2. This analog input V_2, is the voltage to be digitized.

If the analog input voltage V_2 exceeds the resistive network's voltage V_1, the output of U_2 goes HIGH, and the up-down counter counts up, bringing the network's output voltage V_1 up, in steps, to the analog input V_2. On the other hand, if V_2 is less than or decreases below the voltage V_1, the output of U_2 goes LOW, and the up-down counter counts down, bringing voltage V_1 in line with V_2 again.

Generally then, this system has feedback which keeps the voltage output of the resistive network approximately equal to the analog input voltage V_2. Thus, the output of the counter is, or seeks to become, a digital equivalent of the analog input V_2. Through not shown in Fig. 10-18, the counter's outputs can be used to drive a digital readout via an IC latch. The diodes prevent excessive differential inputs to U_2. The zeners are selected to clip the comparator's output to levels compatible with the up-down counter.

More detailed equivalents of the analog switches are shown in Fig. 10-19. They serve to apply either a regulated reference voltage V_{ref} or ground, to represent logic 1 or 0, to each resistor R. This improves the accuracy of the ladder's output.

10.6 ABSOLUTE VALUE CIRCUITS

The output signal V_o of the circuit in Fig. 10-20 can swing positively only, regardless of the polarity of the input signal V_s. Though the input V_s swings through positive and negative values, the output V_o will change proportionally but will vary between zero and positive values. An input V_s and a resulting output V_o for this circuit are shown in Fig. 10-20b and c.

When V_s is positive, diode D_2 is forward-biased, and due to the divider action of the equal resistors R_a and R_b, the signal at input 2 is $V_s/2$. Since D_1 is reverse-biased, the equal resistors R_F and R_1 form a simple feedback network of a noninverting amplifier, and therefore the output is

$$V_o = A_v \frac{V_s}{2} \cong \left(\frac{R_F}{R_1} + 1\right) \frac{V_s}{2} \cong V_s$$

On the other hand, when V_s is negative, D_1 is forward-biased and D_2 is reverse-biased. Since the left side of R_1 is grounded and now the right side is virtually grounded, it draws negligible current and drops out of consideration as far as circuit gain is concerned. Now the output is

$$V_o = A_v V_s \cong -(R_F/R_1')V_s \cong -V_s$$

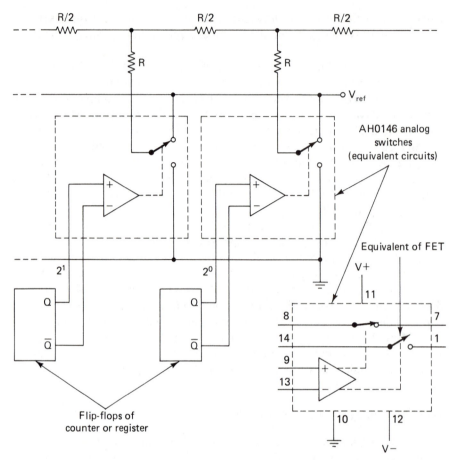

Figure 10-19 Analog switches used to apply either the reference voltage V_{ref} or ground to the inputs of the ladder.

Therefore, if resistors R_1, R_1', R_F, R_a, and R_b are all equal, this circuit's gain is either unity or minus unity, depending on the polarity of the input V_s.

The circuit of Fig. 10-21a is an absolute value detector with high input impedance. Its equivalent circuit on positive alternations of the input signal V_s is shown in (b) of this figure. On negative alternations, the equivalent becomes circuit (c). Note the gain equations inside the brackets at the outputs of each equivalent circuit.

In Fig. 10-21a, we see that the input signal V_s has two amplifying paths; one via the noninverting input of U1 and the inverting input of U2. The other path is via the noninverting input of U2 alone. In Fig. 10-21b, the first term of the gain equation represents the gain of V_s via amplifiers U1 and U2. The

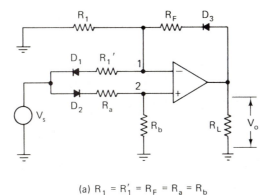

(a) $R_1 = R_1' = R_F = R_a = R_b$

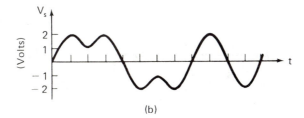

(b)

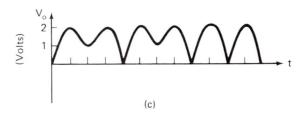

(c)

Figure 10-20   (a) Absolute value output circuit. With input waveform (b) its output has waveform (c).

second term represents the gain of U2 alone. The superposition of these gain paths is the total gain which is unity (1). Note that on positive alternations of V_s, the far-left resistor R and diode D_2 drop out as far as circuit gain is concerned.

As in Fig. 10-21b, the first term inside the brackets in Fig. 10-21c represents the gain via U1 and U2. The second term is the gain of U2. Overall, as shown, the gain is -1 on negative alternations. In this case, diode D_1 drops out of amplifier action.

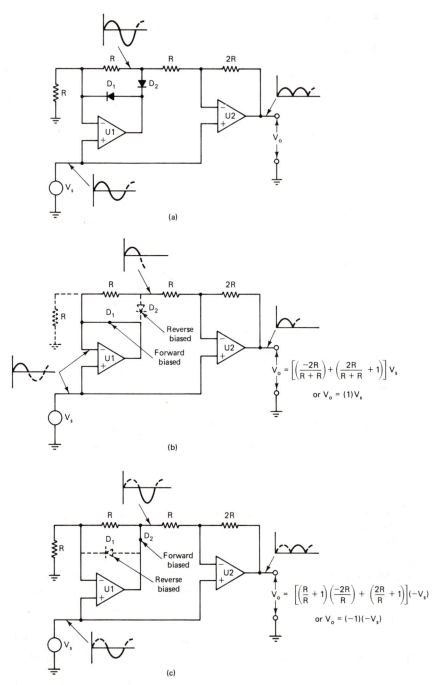

Figure 10-21   (a) Absolute value circuit with high input impedance, (b) its equivalent on positive alternations of V_s, and (c) its equivalent on negative alternations of V_s.

10.7 THE OP AMP AS A SAMPLE-AND-HOLD CIRCUIT

The function of the sample-and-hold circuit is somewhat explained by its name. It usually is used to read (sample) an input signal V_s and hold its instantaneous value for a period of time t_H. An Op Amp sample-and-hold circuit is shown in Fig. 10-22. The MOSFET* serves as a switch that is effectively opened or closed by the presence or absence of a control voltage V_c, on its gate G. A positive pulse V_c applied to the gate G of the enhancement-mode MOSFET causes it to become conductive between its drain D and source S leads. This allows the signal V_s to charge the capacitor C. In fact, the voltage across the capacitor essentially follows V_s when V_c is applied. The time periods when pulses V_c are applied are called *sample* periods t_S. The times when gating pulses V_c are not applied are called *hold* periods t_H. During hold periods t_H, the MOSFET is nonconductive, and the charge in and the voltage across the capacitor C are held constant. The Op Amp's output is usually read or observed during such times.

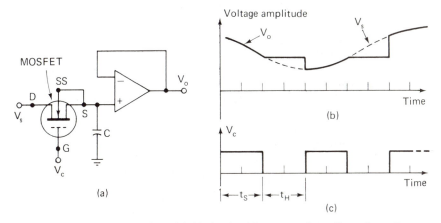

Figure 10-22 (a) A sample-and-hold circuit with output voltage V_o vs. input V_s (b) resulting from the applied control voltage V_c waveform (c).

The sample-and-hold circuit can be used to provide a steady voltage into a device that cannot process a varying signal. An analog-to-digital (A/D) converter might be such a device. Waveforms V_o and V_s in Fig. 10-22b are output and input waveforms of the sample-and-hold circuit when the control voltage V_c has the waveform shown in Fig. 10-22c.

If the times t_S and t_H are short compared with the time it takes the signal V_s to significantly change, successive readings of V_o at instances in t_H give a close approximation of the input waveform. Of course, since the capacitor's function is to hold a constant voltage for periods of time, it

* The MOSFET is a Metal-Oxide Semiconductor Field Effect Transistor.

should be a low-leakage type, preferably one with a polycarbonate, polyethylene, or teflon dielectric.

PROBLEMS

10-1. Sketch the waveform of V_o in the circuit of Fig. 10-1 if V_z of each zener is 6 V, $R_F = 100$ kΩ, $R_1 = 10$ kΩ, and the input signal V_s is sinusoidal with a peak of 1 V. Assume that the Op Amp has negligible offset and that the forward drops of both diodes are negligible.

10-2. In the circuit described in Problem 10-1, sketch the waveform of V_o if D_2 is replaced with a zener whose $V_z = 3$ V.

10-3. Sketch the waveform of V_o in the circuit of Fig. 10-2 if V_z of the zener is 4 V, $R_F = 220$ KΩ, $R_1 = 5$ kΩ, and the input signal V_s is sinusoidal with a peak of 250 mV. Assume that the Op Amp is nulled and that the forward drops across both diodes are negligible.

10-4. Sketch the waveform of V_o in the circuit described in Prob. 10-3 if the 5-kΩ resistor is replaced with 10 kΩ.

10-5. In the circuit described in Prob. 10-3, sketch its output V_o waveform if its diode D_2 is repalced with a short.

10-6. Referring again to the circuit described in Prob. 10-3, sketch the waveform of V_o if either D_1 or D_2 becomes an open.

10-7. If the zener diode D in the circuit of Fig. 10-4 has a $V_z = 10$ V and the input V_s is sinusoidal with a 5-mV peak, what maximum ratio of R_F/R_1 can we use and still prevent clipping of the output signal's positive alternations?

10-8. Sketch the waveform of V_o that we could expect at the output of the circuit in Fig. 10-5 if its zener's $V_z = 8$ V, $R_F = 220$ kΩ, $R_1 = 1.1$ kΩ, and if V_s is sinusoidal with a peak of 80 mV.

10-9. The Op Amps in Fig. 10-23 have output limits of +15 V and − 15 V. With the

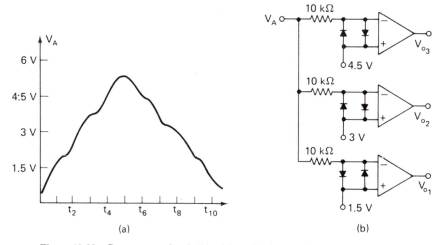

Figure 10-23 Comparator circuit (b) with analog input voltage V_A (a) applied.

input voltage V_A shown, what are the output voltages V_{o_1}, V_{o_2}, and V_{o_3} at instants (a) t_2, (b) t_3, and (c) t_5? A_{VOL} of each Op Amp is extremely large.

10-10. In the circuit and input described in Problem 10-9, what are the outputs V_{o_1}, V_{o_2}, and V_{o_3} at times (a) t_7, (b) t_9, and (c) t_{10}? Assume that the open-loop gain A_{VOL} of each Op Amp is extremely large.

10-11. Sketch the V_o and V_L waveforms of the circuit in Fig. 10-24b, where the waveform in Fig. 10-24a is the input signal V_s, and $V_R = 0$ V to ground.

10-12. Sketch the V_o and V_L waveforms of the circuit in Fig. 10-24b, where Fig. 10-24a shows the input signal V_s, and $V_R = +1.4$ V dc to ground.

10-13. If the Op Amp of Fig. 10-10 has output limits of ± 12 V, $R_a = 1$ kΩ and $R_b = 27$ kΩ, (a) sketch the resulting transfer function. Show the values of V_a' and V_a on the V_A scale. (b) What is the noise margin in volts?

10-14. If the Op Amp of Fig. 10-10 has output limits of ± 12 V, $R_a = 1$ kΩ and $R_b = 15$ kΩ, (a) sketch the circuit's transfer function. (b) What is the noise margin in volts?

10-15. If the signal of Fig. 10-11 is applied to the input of the circuit described in Prob. 10-13, sketch the resulting output waveform.

10-16. If the signal of Fig. 10-11 is applied to the input of the circuit described in Prob. 10-14, sketch the resulting output waveform.

10-17. Suppose that a signal source that has an open-circuit output voltage that is a 3-V peak sinewave and 1 kΩ of internal resistance, is used to drive the input of the circuit in Fig. 10-20. If $R_1 = R_1' = R_F = R_b = 1$ kΩ, sketch the resulting output waveform.

10-18. Suppose that a signal source that has 10 kΩ of internal resistance and a 3-V

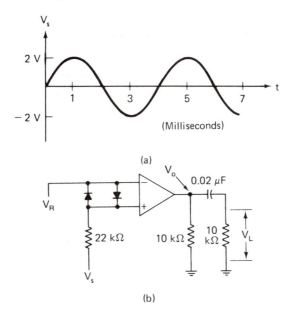

(a)

(b)

Figure 10-24

peak sinewave open-circuit output, is used to drive the input of the circuit of Fig. 10-21a. If $R = 2.2$ kΩ, sketch the resulting output waveform.

10-19. In the circuit of Fig. 10-13, if $R_1 = 40$ kΩ, $R_2 = 10$ kΩ, and $V_{ref} = -1$ V, what are the (a) upper and (b) lower threshold voltages? $V_{o(max)} = \pm 10$ V.

10-20. If $R_1 = 10$ kΩ, $R_2 = 1$ kΩ, and $V_{ref} = +2$ V in the circuit of Fig. 10-13, what are the (a) upper and (b) lower threshold voltages? $V_{o(max)} = \pm 12$ V.

10-21. In the circuit described in Prob. 10-19, if the input signal V_s is a 100-Hz sinewave with a peak of 6 V, sketch the output waveform and indicate the approximate width of its positive and negative alternations in milliseconds.

10-22. In the circuit described in Prob. 10-20, if V_s is a 400-Hz sinewave with a peak of 10 V, sketch the output waveform and indicate the approximate width of its positive and negative alternations in milliseconds.

10-23. What is the voltage V_o in the circuit of Fig. 10-25 if the switches 2^0, 2^1, 2^2, and 2^3, are in the 1, 1, 0, and 0 positions, respectively?

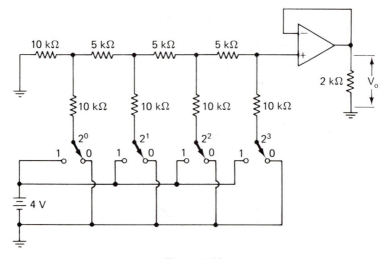

Figure 10-25

10-24. What is V_o in the circuit of Fig. 10-25 for each of the following sets of switch positions?

	2^0	2^1	2^2	2^3	V_o
(a)	1	0	0	0	
(b)	1	1	0	0	
(c)	0	0	1	0	

11

OP AMP
SIGNAL GENERATORS

Op Amps can be wired to serve as signal generators capable of a variety of output waveforms. Squarewaves, triangular waves, sawtooth waves, and sinewaves are readily available, to name the more useful waveforms. In this chapter, we will see a few basic Op Amp signal generators and methods of selecting their externally wired components.

11.1 A SQUAREWAVE GENERATOR

A simple Op Amp squarewave generator is shown in Fig. 11-1. Its output repetitively swings between positive saturation $+V_{o(max)}$ and negative saturation $-V_{o(max)}$, resulting in the squarewave output shown. The time period T of each cycle is determined by the time constant of the components R and C and by the ratio R_a/R_b. This circuit's operation can be analyzed as follows: At the instant the dc supply voltages, $+V$ and $-V$, are applied. zero volts of the initially uncharged capacitor C is applied to the inverting input 1, that is, input 1 is initially grounded. At the same instant, however, a small positive or negative voltage V_b appears across R_b, and this voltage is applied to the noninverting input 2. Voltage V_b initially appears because a positive or negative output offset voltage V_{oo} exists, even if no differential input voltage is applied to inputs 1 and 2. Thus, the resistors R_a and R_b form a voltage divider, and a fraction of the Op Amp's output voltage is dropped across R_b. Since the inverting input 1 is initially grounded through the uncharged capacitor C, all of the voltage V_b initially appears across the inputs 1 and 2.

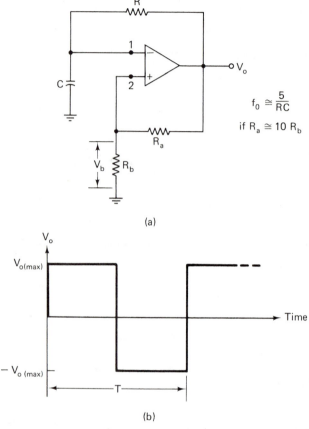

$$f_0 \cong \frac{5}{RC}$$

if $R_a \cong 10\ R_b$

(a)

(b)

Figure 11-1 (a) Squarewave generator, and (b) output waveform of the squarewave generator at low frequencies f where $f = 1/T$.

Even if V_b is small, it will start to drive the Op Amp into saturation. If the output offset V_{oo} is positive, the voltage V_b at the noninverting input 2 is positive. This V_b is initially amplified by the Op Amp's open-loop gain A_{VOL} and drives the output to its limit $V_{o(max)}$, that is, to positive saturation. The rise to $V_{o(max)}$ is at the slew rate of the Op Amp. With the Op Amp saturated, the capacitor charges through resistor R. If the resistor R and capacitor C formed a simple RC circuit, the capacitor's voltage V_c would eventually rise to $V_{o(max)}$. In this case, however, voltage V_c can rise only to a value slightly more positive than V_b. That is, as V_c rises and becomes a little more positive than V_b, the inverting input 1 becomes more positive than the noninverting input 2, and this drives the output to its negative limit $-V_{o(max)}$. See Fig. 11-2. After the Op Amp's output saturates at $-V_{o(max)}$, a fraction of this voltage is dropped across R_b. Thus input 2 becomes much more negative than input 1 and holds the Op Amp in negative saturation, at least for a while. The

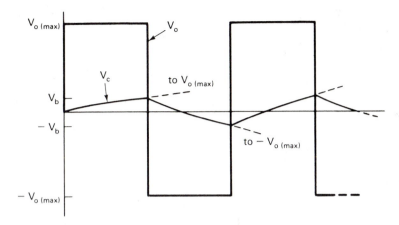

Figure 11-2   Typical output voltage V_o and capacitor voltage V_c waveforms of the squarewave generator.

capacitor C now proceeds to discharge and recharge in the negative direction as shown in Fig. 11-2. Now when the capacitor's voltage becomes more negative than $-V_b$, the inverting input 1 becomes more negative than input 2, and the output is driven back to $+V_{o(max)}$ to start another cycle.

The *sum* of the resistances R_a and R_b is not critical. It can be selected to be in a broad range, say 10 kΩ to 1 MΩ. A change in their *ratio*, however, does affect the circuit's output frequency f. The period of each cycle is

$$T = 2\,RC\,\log_e\left(1 + \frac{2R_b}{R_a}\right)^* \tag{11-1}$$

and the number of cycles per second

$$f = 1/T \tag{11-2}$$

The smaller the RC product, the faster the capacitor C charges to the voltage across R_b, and the higher the output frequency. Therefore, the resistor R in Fig. 11-1 can be a frequency-selecting potentiometer.

This squarewave generator's output frequency is limited by the slew rate of the Op Amp. If we attempt to operate it at relatively high frequencies, the period of each cycle T is not much larger than the time it takes the output to rise to $V_{o(max)}$ from $-V_{o(max)}$, and vice versa. This causes the generator's output to become trapezoidal or even triangular at higher frequencies.

The peak-to-peak output capability of this squarewave generator can be reduced by means of reduced dc supply voltages or with back-to-back zeners as shown in Fig. 11-3. The zeners will reduce ringing† as well.

---

* Use the ln key on your calculator.

† Ringing refers to an output voltage that oscillates about its eventual steady state after a sudden change.

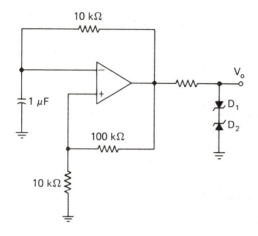

Figure 11-3

11.2 A TRIANGULAR-WAVE GENERATOR

In Section 8-3 we learned that an integrator's output waveform is triangular if its input is a squarewave. See Fig. 8-12a. Apparently, then, an integrator following a squarewave generator, such as in Fig. 11-4, serves as a triangular-wave generator. Since changes in resistance R change the frequency of the squarewave generator's output, the output frequency of the integrator changes. Therefore, if the resistance R is increased or decreased, the frequency of the triangular wave will decrease or increase, respectively. The amplitude of the triangular wave can be controlled somewhat by resistance R_1. Larger or smaller values of R_1 will reduce or increase the output amplitude of the integrator. As with the simple squarewave generator, the output frequency is limited by the Op Amps' slew rates.

11.3 A SAWTOOTH GENERATOR

A sawtooth waveform differs from the triangular waveform in its unequal rise and fall times. The sawtooth may rise positively many times faster than it falls negatively, or vice versa. The circuit in Fig. 11-5 is a sawtooth generator. The first stage is called a *threshold detector*. Its output V_o will swing from its $V_{o(\max)}$ to $-V_{o(\max)}$ when the decreasing output V_o' of the integrator becomes negative enough to pull point x slightly negative with respect to ground. Note the waveforms in Fig. 11-5b. On the other hand, the threshold detector's output swings from $-V_{o(\max)}$ to $+V_{o(\max)}$ when the rising voltage V_o' lifts point x slightly positive with respect to ground. The decrease of V_o' in the negative direction takes longer than its rise in the positive direction because the rate at which the capacitor C charge changes as the polarity of V_o changes. That is, when V_o is saturated negatively, the

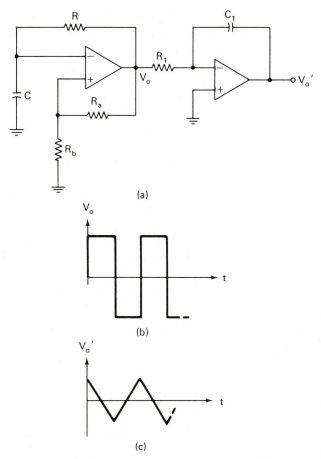

Figure 11-4   (a) Triangular-wave generator, (b) output waveform of the first Op Amp, and (c) output of the second Op Amp.

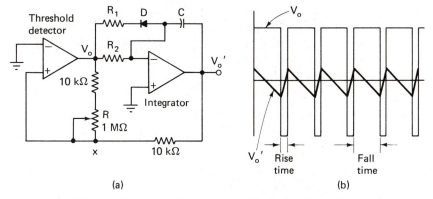

Figure 11-5   (a) The sawtooth generator and (b) its waveform if $R_2 > R_1$.

capacitor C charges mainly through R_1 and the forward-biased diode D. The time constant R_1C is relatively short if R_1 is made significantly smaller than R_2. When V_o is saturated positively, the capacitor C charges through R_2 more slowly because the time constant R_2C is relatively long. The values of R_1 and R_2 largely dictate the frequency of the output V'_o, while their ratio R_1/R_2 determines the ratio of the rise and fall times. If we reverse the diode D, the rise time of V'_o becomes larger than the fall time.

11.4 VARIABLE-FREQUENCY SIGNAL GENERATORS

The circuit in Fig. 11-6 is an extension of the triangular-wave generator in Fig. 11-4. A broad range of output frequencies can be selected by the six-position switch shown. Each higher switch position increases the output frequency by ten. Of course, if we intend to use the high-frequency positions 5 and 6, we would select a suitable high-frequency Op Amp with a high slew rate. The 25-kΩ potentiometer is a fine frequency control. The 50-kΩ symmetry control enables us to vary the ratio of the time of each positive

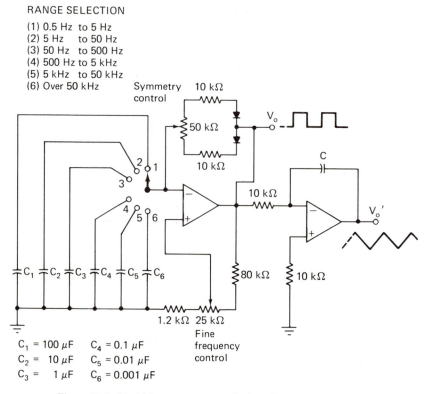

RANGE SELECTION

(1) 0.5 Hz to 5 Hz
(2) 5 Hz   to 50 Hz
(3) 50 Hz to 500 Hz
(4) 500 Hz to 5 kHz
(5) 5 kHz to 50 kHz
(6) Over 50 kHz

$C_1 = 100\ \mu F$ $C_4 = 0.1\ \mu F$
$C_2 = 10\ \mu F$ $C_5 = 0.01\ \mu F$
$C_3 = 1\ \mu F$ $C_6 = 0.001\ \mu F$

Figure 11-6 Variable squarewave and triangular-wave generator.

alternation to the time of each negative alternation of the waveforms V_o and V_o'. Thus the output V_o' can be changed from a triangular to a sawtooth waveform by the symmetry control. Capacitor C largely determines the amplitude of V_o'. Generally, the capacitor C must be larger with lower output frequencies. If C is too small, the output V_o' becomes clipped because the output Op Amp saturates on each alternation. If C is too large, the amplitude of V_o' is very small, especially at higher frequencies.

11.5 THE WIEN BRIDGE OSCILLATOR

A Wien bridge oscillator is shown in Fig. 11-7. Its output is a sinewave. Note that the Op Amp's output is across a network consisting of R_1, C_1, R_2, and C_2 which is the frequency-sensitive portion of a Wien bridge. The signal at point x of the Wien bridge network, which is the input signal on the noninverting input, is in phase with the signal V_o at a particular frequency f_c. This frequency is

$$f_c = \frac{1}{2\pi R_1 C_1} \tag{11-3}$$

if

$$R_1 = R_2$$

and

$$C_1 = C_2$$

The feedback signal at x leads V_o at frequencies below f_c and lags V_o at frequencies above f_c. Of course then, maximum in-phase feedback occurs at

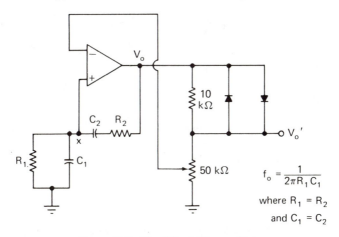

Figure 11-7 The Wien bridge oscillator.

f_c, which therefore is the output frequency of the oscillator. Adjustment of the 50-kΩ potentiometer controls the amount of negative feedback to the inverting input and the amplitude of the output V_o'. The diodes prevent excessive amplitude of negative feedback.

11.6 THE TWIN-T OSCILLATOR

The arrangement of resistors and capacitors, shown in Fig. 11-8a is called a passive twin-T notch filter. With sinewave voltage input V_{in}, the resistors R_1 and capacitor C_1 (upper T) cause a voltage component at X to lag the input V_{in}. If the frequency of V_{in} is increased, the lagging phase angle between the voltage component at X and V_{in} increases while the amplitude of this voltage component decreases. On the other hand, the capacitors C_2 and resistor R_2 (lower T) cause a voltage component at x to lead V_{in}. If the frequency of V_{in} is increased, this leading phase angle decreases while the amplitude of the voltage component at X increases. The superposition of these two voltage components; one out of the upper T and the other out of the lower T is the output voltage V_o. At what is called *the center frequency* f_c, the two voltage components are out of phase and have equal amplitudes. At f_c, therefore the two voltage components tend to cancel each other and cause the amplitude of V_o to become minimum. The resulting V_o vs. f curve is as shown in Fig. 11-8b. At the center frequency

$$f_c = \frac{1}{2\,\pi R_1 C_2} \tag{11-4}$$

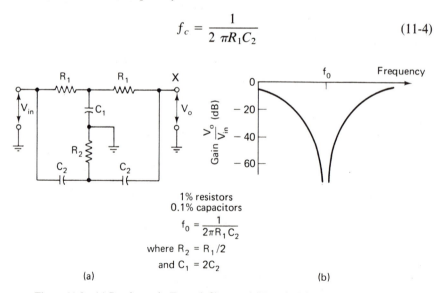

1% resistors
0.1% capacitors

$$f_0 = \frac{1}{2\pi R_1 C_2}$$

where $R_2 = R_1/2$
and $C_1 = 2C_2$

(a)   (b)

Figure 11-8   (a) Passive twin-T notch filter, and (b) typical frequency response of the twin-T notch filter.

where

$$R_1 = 2R_2$$

and

$$C_1 = 2C_2$$

A twin-T sinewave oscillator is shown in Fig. 11-9. It has two feedback loops. The loop consisting of the 10 kΩ resistor and the 50 kΩ POT causes positive (regenerative) feedback. Positive feedback tends to make the circuit unstable and to go into oscillation. The loop consisting of the twin-T filter is in the negative (degenerative) feedback loop. This filter admits minimum negative feedback at its center frequency f_c causing the circuit to go into oscillation at that frequency.

The component values in the twin-T oscillator are critical; 0.1% resistors and 1% capacitors are recommended. The 50-kΩ POT enables us to optimize the waveform and amplitude of V_o.

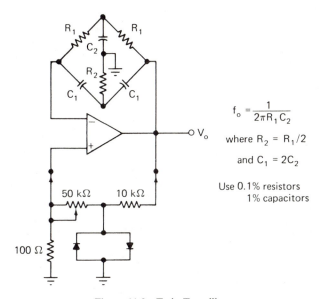

$$f_o = \frac{1}{2\pi R_1 C_2}$$

where $R_2 = R_1/2$

and $C_1 = 2C_2$

Use 0.1% resistors
1% capacitors

Figure 11-9   Twin-T oscillator.

PROBLEMS

11-1. In the circuit of Fig. 11-1, if $R_a = 100$ kΩ, $R_b = 10$ kΩ, $R = 20$ kΩ, and $C = 1$ μF, what is its approximate output frequency f_o?

11-2. If $R_a = 200$ kΩ, $R_b = 20$ kΩ, and $C = 1$ μF, what value of resistance R should we use in the circuit of Fig. 11-1 if we need approximately a 100-Hz output?

11-3. In the circuit described in Prob. 11-1, about what maximum peak-to-peak value of output signal voltage can we expect if the Op Amp is a type 741 and the dc supply voltages are $+15$ V and -15 V?

11-4. If the Op Amp in the circuit of Fig. 11-3 is a type 741 operated with dc supply voltages of $+20$ V and -20 V, and if each zener has a zener voltage $V_z = 6.3$ V, what approximate peak-to-peak output signal voltage can we expect?

11-5. If C_1 is too small in the circuit of Fig. 11-4, what effect might it have on the output waveform V_o'?

11-6. If R_1 is too large in the circuit of Fig. 11-4, what effect will it have on the amplitude of the output waveform V_o'?

11-7. Sketch the approximate waveform of the output V_o' that we could expect from the circuit in Fig. 11-5 if $R_1 = 1$ kΩ and $R_2 = 20$ kΩ?

11-8. Sketch the approximate waveform of the output V_o' that we could expect from the circuit in Fig. 11-5 if the diode becomes open?

11-9. Select component values for the circuit of Fig. 11-7 so that its output frequency is about 500 Hz. Use $C_1 = 0.2$ μF.

SINGLE-SUPPLY DEVICES

The Op Amp's ability to provide an output voltage swing in both positive and negative directions is not always necessary or desirable. In such cases, the Op Amp's need of positive and negative supply voltages is a costly inconvenience. Devices made to work with one supply are available. Popular types, called *quads*, and a few of their applications are discussed here. The term *quad* describes the fact that there are four separate devices in a single IC package. Single-supply quads are not one-for-one replacements for Op Amps. As we will see, they require specialized external wiring.

12.1 THE CURRENT-DIFFERENCING AMPLIFIER

An economical and easy to use single-supply device is shown in Fig. 12-1. It is a package containing four *current-differencing* (CD) amplifiers*. The circuit of each CD amplifier is fundamentally as shown in Fig. 12-2. Q_1 is the active transistor of a common-emitter amplifier and Q_2 serves as its load resistance. The signal from the collector of Q_1 drives the base of Q_3 which operates as an emitter follower with its load being transistor Q_4. This arrangement is capable of gains over 60 dB, open loop. Transistors Q_2 and Q_4 are biased to behave as constant-current sources and therefore have very high dynamic resistances.

---

* Typical manufacturers' part numbers for current-differencing amplifiers are LM2900 and LM3900 (National Semiconductor) and MC3401 (Motorola).

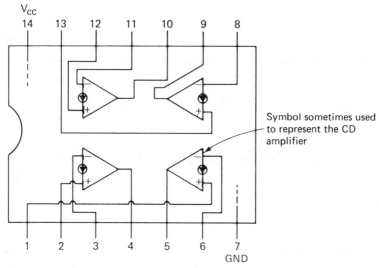

Figure 12-1 The quad amplifier package contains four current-differencing amplifiers and a bias supply.

Note the output terminal of this circuit of Fig. 12-2. When the base of Q_3 is driven hard, causing this transistor to conduct very well, nearly all of the dc source voltage V_{CC} appears across Q_4 and is the output V_o. On the other hand, with little or no drive on Q_3, it is cut off and nearly all of the V_{CC}

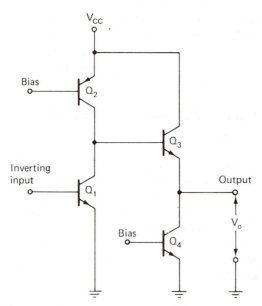

Figure 12-2 Amplifying section of a current-differencing amplifier. Q_2 and Q_4 are biased to behave as high ac resistances (current sources).

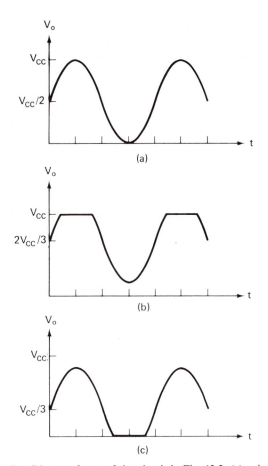

Figure 12-3 Possible waveforms of the circuit in Fig. 12-2; (a) quiescent V_o at $V_{CC/2}$ gives maximum peak-to-peak output capability; (b) quiescent V_0 higher than $V_{CC/2}$ causes clipping of positive peaks; (c) quiescent V_o lower than $V_{CC/2}$ causes clipping of negative peaks.

voltage appears across its collector and emitter terminals. In this case, the output voltage V_o, which is the voltage across Q_4, is about 0 V. Therefore, for linear applications and relatively large output signal capability, this amplifier should be biased so that the quiescent dc output voltage V_o is about half the V_{CC} voltage. As shown in Fig. 12-3, more clipping occurs with larger output signals if the output is not biased at $V_{CC}/2$.

12.2 FEEDBACK WITH THE CD AMPLIFIER

In the preceding chapters, we reduced the effective gain of an Op Amp by placing a feedback resistor across its output and inverting input. We can take the same approach to control the gain of the CD amplifier circuit. Use of a feedback resistor R_F, however, presents a problem. Since the CD amplifier, intended for linear operation, is biased so that its output is above ground potential, an average dc current I_F is forced through R_F as shown in Fig. 12-4. That is, the right end of R_F is above ground by the quiescent value of V_o, while its left end is above ground by the base-to-emitter voltage V_{BE} of Q_1, which typically is just a few tenths of a volt. This causes a dc potential difference across R_F that is $V_o - V_{BE}$; therefore, the dc current through this feedback resistor is

$$I_F = \frac{V_o - V_{BE}}{R_F} \tag{12-1}$$

If the dc supply voltage V_{CC} is significantly larger than V_{BE}, and if the quiescent value of V_o is to be half of the V_{CC} voltage, for good peak-to-peak output signal capability, Eq. (12-1) can be modified to

$$I_F \cong \frac{V_{CC}/2}{R_F} = \frac{V_{CC}}{2R_F} \tag{12-2}$$

In a practical case, this feedback current I_F is much too large to properly bias the base of Q_1. In fact, if most of I_F is not diverted from the base of Q_1,

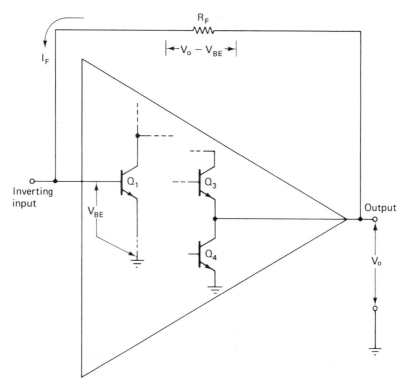

Figure 12-4 The CD amplifier with a feedback resistor R_F.

the amplifier will saturate. As shown in Fig. 12-5, additional circuitry, including transistor Q_5, provides this necessary shunt path to ground. The conductivity of Q_5 is controlled by current I_2, and during linear operation the average values of currents I_2 and I_F are equal. In fact, because I_F seeks to be equal to (reflect) I_2, the I_2 current is called a *mirror current*. As users of CD amplifiers, we select a value of I_2 between 10 μA and 500 μA and establish it by a proper choice of resistor R_2. Note that as the inverting input is at about ground potential through the forward-biased base-emitter junction of Q_1, so likewise the noninverting input is grounded through the base-emitter of Q_5. Thus, by Ohm's law,

$$I_2 = \frac{V_{CC} - V_{BE}}{R_2} \tag{12-3a}$$

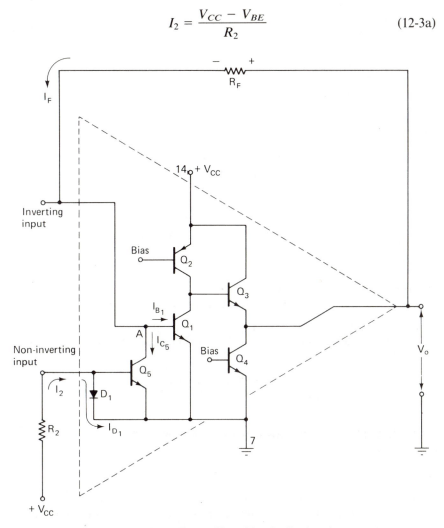

Figure 12-5   A practical CD amplifier with a feedback resistor R_F.

or

$$I_2 \cong \frac{V_{CC}}{R_2} \qquad\qquad (12\text{-}3b)$$

If the output transistor Q_3 is neither saturated nor cutoff, we can view the CD amplifier as a device that adjusts its output voltage V_o such as to remove any differences in its input currents I_F and I_2. Should I_F be reduced, say by an increase in R_F, the output V_o will rise proportionally to bring I_F back to a value equal to I_2. If, on the other hand, I_2 is decreased by an increase in R_2, the output V_o will decrease to bring I_F down to the new value of I_2. Since currents I_F and I_2 seek to be equal, make

$$R_F = R_2/2 \qquad\qquad (12\text{-}4)$$

for a quiescent $V_o = V_{CC}/2$.

Example 12-1

If a CD amplifier is working with a 12-V supply, solve for the values of R_2 and R_F needed for a mirror current of 10 μA and a $V_o = 6$ V.

Answers.   Since we know that $V_{CC} = 12$ V, we can rearrange Eq. 12-3b to find that

$$R_2 \cong \frac{V_{CC}}{I_2} = \frac{12 \text{ V}}{10 \text{ }\mu\text{A}} = 1.2 \text{ M}\Omega$$

By Eq. 12-4 then

$$R_F = \frac{R_2}{2} = \frac{1.2 \text{ M}\Omega}{2} = 600 \text{ k}\Omega$$

Example 12-2

Referring to the previous example, if due to availability, we must use an $R_F = 560$ kΩ instead of 600 kΩ, what will the quiescent voltage V_o be? Assume that $R_2 = 1.2$ MΩ.

Answer.   With $R_2 = 1.2$ MΩ, the mirror current $I_2 = 10$ μA and V_o seeks a level that will make $I_F = 10$ μA. Since the left end of R_F is grounded, V_o is the drop across R_F and in this case will be

$$V_o \cong R_F I_F = 560 \text{ k}\Omega \text{ } (10 \text{ }\mu\text{A}) = 5.6 \text{ V}$$

12.3 THE INVERTING CURRENT-DIFFERENCING AMPLIFIER

A simple inverting circuit using a CD amplifier is shown in Fig. 12-6. It can be used to amplify ac but not dc signals because of the input coupling capacitor C_1. This capacitor is needed to prevent any part of the dc current I_F from flowing to ground through R_1 and the signal source V_S.

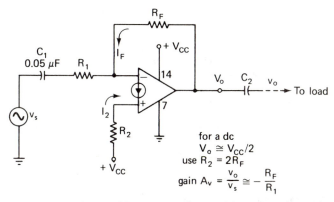

Figure 12-6   The CD amplifier connected to work in an inverting mode.

From the ac signal's point of view, this circuit can be analyzed much as the conventional Op Amps were in preceding chapters, resulting in the same gain equation. At frequencies that see the reactances of the coupling capacitors C_1 and C_2 as negligible (see Appendix L), this inverting circuit's gain is

$$A_v = \frac{V_o}{V_s} \cong -\frac{R_F}{R_1} \tag{12-5}$$

The feedback resistor R_F is selected so that the feedback current is about equal to the manufacturer's recommended mirror current. After the required R_F is established, the value of R_1 is selected to provide the desired stage gain. Then R_2 is selected to inject a mirror current into the noninverting input that is equal to the feedback current I_F.

Example 12-3

Select resistance values of R_F, R_1, and R_2 for the circuit in Fig. 12-6, using a CD amplifier, to provide a voltage gain of 100. Use a mirror current of 10 μA and a dc source $V_{CC} = +20$ V.

Answer.   Since $I_F = V_{CC}/2R_F$ according to Eq. (12-2), and since it is to be equal to the mirror current, we can solve for R_F and show that

$$R_F = \frac{V_{CC}}{2I_F} = \frac{20 \text{ V}}{2(10 \text{ }\mu\text{A})} = 1 \text{ M}\Omega$$

Therefore, for a gain of 100, the input resistor R_1 must be smaller by the factor 100. By rearranging Eq. (12-5), we can show that

$$R_1 \cong \frac{R_F}{-A_v} = \frac{1 \text{ M}\Omega}{100} = 10 \text{ k}\Omega$$

Then for the required mirror current I_2,

$$R_2 \cong 2R_F = 2(1 \text{ M}\Omega) = 2 \text{ M}\Omega \tag{12-4}$$

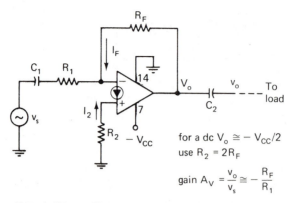

Figure 12-7   A CD amplifier connected to work with a negative supply.

The CD amplifier can also be operated with a single negative dc supply as shown in Fig. 12-7. In this case, the ground pin of the package is connected to the $-V_{CC}$ source, and the $+V_{CC}$ pin of the package is grounded or connected to the common of the $-V_{CC}$ source. In fact, all previously grounded points of the IC are placed at $-V_{CC}$ potential, and all points previously at $+V_{CC}$ are grounded. The external component selection for this circuit is made in the same way as for the previous circuit (Fig. 12-6). In this case, both inputs are at about $-V_{CC}$ potential, instead of grounded, through the base-emitter junctions of transistors Q_1 and Q_5.

12.4 THE NONINVERTING CURRENT-DIFFERENCING AMPLIFIER

The circuit in Fig. 12-8 is connected to work as a noninverting CD amplifier using a single positive dc source voltage $+V_{CC}$. The bias and gain determining components for this noninverting circuit are selected exactly as they are

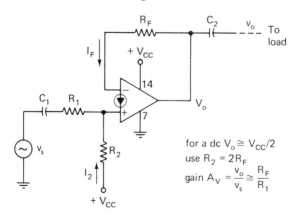

Figure 12-8   The CD amplifier connected to work as a noninverting amplifier.

for the inverting types. Of course, the negative sign in the gain equation is not used because of the in-phase relationship of the output and input signal voltages in this case.

Example 12-4

Select values of R_F, R_1, and R_2 for the noninverting circuit in Fig. 12-8, where the voltage gain is to be about 200 and the recommended mirror current is 50 μA. The dc supply $V_{CC} = 15$ V.

Answer.   Since the feedback and mirror currents are to be equal for linear operation, we first solve for R_F by rearranging Eq. (12-2). Thus,

$$R_F = \frac{V_{CC}}{2I_F} = \frac{15 \text{ V}}{2(50 \text{ }\mu\text{A})} = 150 \text{ k}\Omega$$

To obtain a gain $A_v = 200$, R_1 must be

$$R_1 \cong \frac{R_F}{A_v} = \frac{150 \text{ k}\Omega}{200} = 750 \text{ }\Omega \qquad (12\text{-}5)$$

and by Eq. (12-4),

$$R_2 \cong 2R_F = 2(150 \text{ k}\Omega) = 300 \text{ k}\Omega$$

12.5 THE CD AMPLIFIER IN DIFFERENTIAL MODE

CD amplifiers can be wired to work in differential mode as shown in Fig. 12-9. With proper selection of components, this circuit has good common-mode rejection (*CMRR*). For a dc quiescent output $V_o = V_{CC}/2$,

$$R_b + R_C = 2R_F \qquad (12\text{-}6)$$

Similar to previous circuits, this circuit's voltage gain is

$$A_v = \frac{v_o}{v_s} \cong -\frac{R_F}{R_1} \qquad (12\text{-}5)$$

and for good *CMRR*

$$R_b = R_F \qquad (12\text{-}7a)$$

and

$$R_a = R_1 \qquad (12\text{-}7b)$$

The capacitor C_3 places the bottom of resistor R_b at ac ground potential and is necessary for good *CMRR*. As in the inverting and noninverting circuits discussed in the previous sections, the value of R_F is dictated by the value of the dc supply V_{CC} and the recommended mirror current.

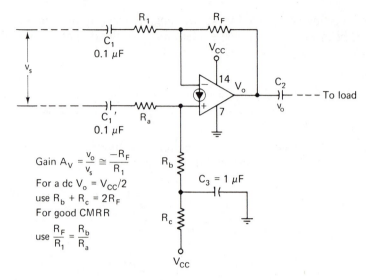

Figure 12-9 The CD amplifier connected to work in a differential mode.

Example 12-5

Select resistor values for the circuit in Fig. 12-9 for good *CMRR* and a voltage gain of 120. The dc supply $V_{CC} = 24$ V and the recommended mirror current is 100 μA.

Answer. First, by Eqs. (12-2) and (12-7a)

$$R_F = R_b = \frac{V_{CC}}{2I_F} = \frac{24 \text{ V}}{2(100 \ \mu\text{A})} = 120 \text{ k}\Omega$$

Now we find that, by Eqs. (12-5) and (12-7b)

$$R_1 = R_a \cong \frac{R_F}{A_v} = \frac{120 \text{ k}\Omega}{120} = 1 \text{ k}\Omega$$

therefore

$$R_b + R_c = 2R_F = 2(120 \text{ k}\Omega) = 240 \text{ k}\Omega \tag{12-6}$$

and finally

$$R_c = 240 \text{ k}\Omega - R_b = 120 \text{ k}\Omega$$

12.6 THE CD AMPLIFIER AS A COMPARATOR

Comparators are used extensively in digital-to-analog converters and in analog-to-digital converters. Their function is to provide fast changes in output voltage levels when their input voltages V_s change through a reference voltage. Two CD comparators are shown in Fig. 12-10. In the circuit shown in Fig. 12-10a, the output $V_0 \cong 0$ V as long as the input voltage

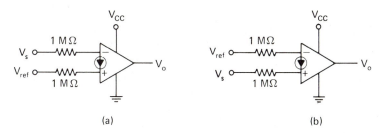

Figure 12-10   CD amplifiers used as comparators: (a) $V_o \cong 0$ when $V_S > V_{ref}$, and $V_o \cong V_{CC}$ when $V_S < V_{ref}$; (b) $V_o \cong 0$ when $V_s < V_{ref}$, and $V_o \cong V_{CC}$ when $V_s > V_{ref}$.

V_s is more positive than the reference voltage V_{ref}. If V_s changes to a value less positive than V_{ref}, the output V_o quickly swings to about the V_{CC} voltage. In the circuit shown in Fig. 12-10b, the output $V_o = 0$ V as long as the input V_s is less positive than the reference V_{ref}. When V_s swings to a value more positive than V_{ref}, the output V_o quickly switches to approximately the V_{CC} voltage.

The CD amplifier makes an efficient comparator because it requires only two current-limiting resistors in its input leads. No feedback resistor is used because the open-loop gain is desirable. The high gain enables us to drive the output V_o between saturation and cutoff with very small changes in the input currents.

12.7 QUAD VOLTAGE COMPARATORS

Several manufacturers make quads called *voltage* comparators that, like CD amplifiers, work with one dc supply. Typical packages are shown in Fig. 12-11. These quads are not intended for linear operation, which means that in typical applications their outputs repetitively switch from one voltage level to another. Note in Fig. 12-12 that each voltage comparator has an open collector output. As we will see, an open collector output usually requires a *pull-up* resistor.

12.7.1 A Limit Detector

An application using two voltage comparators is shown in Fig. 12-13. The LED (light-emitting diode) emits light when the input voltage V_s is between voltages V_1 and V_2. For example, if $V_{CC} = 6$ V and $R_1 = R_2 = R_3$, voltage $V_1 = 2$ V and $V_2 = 4$ V. If V_s is less than 2 V, the output of Amp #1 is pulled to ground potential by its saturated output transistor Q_o (Fig. 12-12). As a result, there is virtually no base bias current for the driver transistor Q causing it to cutoff. The LED, therefore, does not emit light. If V_s is between 2 V and 4 V, the output transistors Q_o of both #1 and #2 Amps

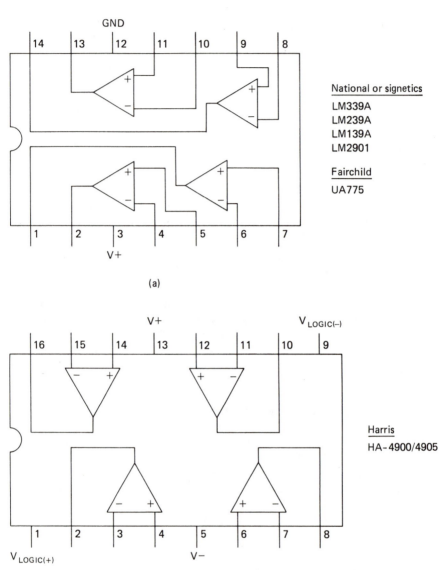

(a)

(b)

Figure 12-11   Quad voltage comparators; (a) 14-pin package, (b) 16-pin package.

cutoff. A resulting base bias current flows by way of R_4 and through the driver Q causing the LED to light. When V_s is above 4 V, the output transistor Q_o of #2 Amp saturates and, as before, steers bias current away from the base of the driver Q. The driver cuts off and the LED remains unlighted. Of course the values of the threshold voltages V_1 and V_2 can be selected by choosing appropriate values of divider resistors R_1, R_2, and R_3.

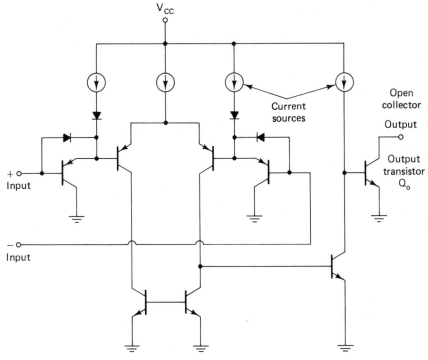

Figure 12-12   Schematic of one voltage comparator.

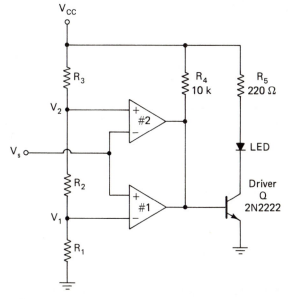

Figure 12-13   Limit detector with voltage comparators and LED.

12.7.2 A Fast Comparator Circuit

The circuit of Fig. 12-14 works as a fast comparator that has hysteresis in its transfer characteristic. It is fast in that its output V_o can switch from one voltage level to another in a relatively short time. As mentioned before, hysteresis helps to prevent switching of the output V_o by noise voltages that might be riding on the input signal V_s.

If the reference voltage V_{ref} is positive and the input voltage is initially zero, the output V_o is at about ground (0 V) potential; see the transfer function of Fig. 12-14. Note that if V_s is increased, the output V_o remains at about 0 V so long as V_s is below the *upper threshold* voltage V_{th}. If an increasing V_s crosses through V_{th}, the comparator's output V_o quickly switches to about the V_{CC} voltage via the 3-kΩ pull-up resistor. Then when V_s is decreased toward 0 V, the output V_o stays at the V_{CC} voltage until V_s becomes less than the *lower threshold* voltage V_{th}'. The upper threshold voltage

$$V_{th} = \frac{V_{ref}(R_1 + R_2)}{R_2} \tag{12-8}$$

and the lower threshold voltage

$$V_{th}' = \frac{V_{ref}(R_1 + R_2) - V_{CC}R_1}{R_2} \tag{12-9}$$

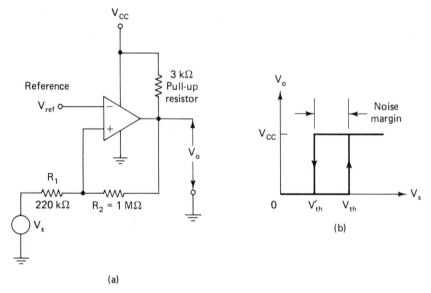

Figure 12-14 (a) Fast comparator circuit using an LM339 and (b) its transfer function.

The noise margin is the difference in Eqs. 12-8 and 12-9; i.e.,

$$\text{Noise margin} = V_{th} - V_{th}' = \frac{V_{CC}R_1}{R_2} \tag{12-10}$$

Example 12-6

Referring to the circuit of Fig. 12-14, if $V_{CC} = 15$ V and $V_{ref} = 5$ V, what are the upper and lower threshold voltages and the noise margin? If V_s is a sinewave with a peak of 7 V, sketch the resulting output V_o.

Answers.   In this case, the upper threshold voltage

$$V_{th} = \frac{5 \text{ V}(220 \text{ k}\Omega + 1 \text{ M}\Omega)}{1 \text{ M}\Omega} = 6.1 \text{ V} \tag{12-8}$$

and the lower threshold voltage

$$V_{th}' = \frac{5 \text{ V}(220 \text{ k}\Omega + 1 \text{ M}\Omega) - 15 \text{ V}(220 \text{ k}\Omega)}{1 \text{ M}\Omega} = 2.8 \text{ V} \tag{12-9}$$

Therefore, the noise margin is 6.1 V − 2.8 V = 3.3 V. With the input signal described, the resulting output V_o is as shown in Fig. 12-15.

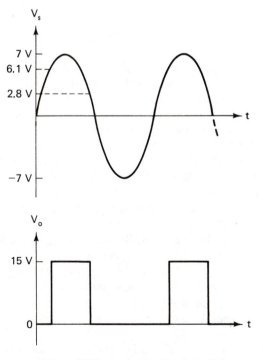

Figure 12-15 Answer to Example 12-6.

REVIEW QUESTIONS

12-1. Name three advantages that CD amplifiers have compared to conventional Op Amps.

12-2. Name three disadvantages that CD amplifiers have compared to conventional Op Amps.

12-3. What type of CD amplifier applications require a mirror current?

12-4. How does the quiescent dc output of the CD amplifier compare to the dc source voltage if the dc feedback current is equal to the mirror current?

12-5. What will happen to the output signal waveform of the CD amplifier if its dc feedback current to the inverting input is not equal to the current in the noninverting input?

12-6. Why does a dc feedback current exist when the CD amplifier is biased for linear operation?

12-7. Can the differential-mode circuit in Fig. 12-9 be used to amplify signals from a thermocouple bridge that is monitoring the temperature of a large furnace? Why?

12-8. What circuit performance will suffer in the circuit of Fig. 12-9 if the capacitor C_3 becomes open?

12-9. If a device has an open-collector output, what kind of external component is usually necessary?

PROBLEMS

12-1. Referring to the inverting amplifier of Fig. 12-6, if $V_{CC} = 12$ V, what values of R_1, R_2, and R_F should we use to establish a mirror current of 40 μA, a voltage gain $A_v = 50$, and a dc quiescent output $V_o = V_{CC}/2$?

12-2. Referring to the amplifier of Fig. 12-6, if $V_{CC} = 9$ V, what values of R_1, R_2 and R_F should we use to establish a mirror current of 50 μA, a gain $A_v = 40$, and a dc quiescent output $V_o = 4.5$ V?

12-3. If in the circuit of Fig. 12-6, $V_{CC} = 12$ V, $R_1 = 3.3$ kΩ, $R_2 = 270$ kΩ, and $R_F = 150$ kΩ, what are its voltage gain A_v and dc quiescent output V_o?

12-4. If in the circuit of Fig. 12-6, $V_{CC} = 9$ V, $R_1 = 2.2$ kΩ, $R_2 = 180$ kΩ, and $R_F = 91$ kΩ, what are its voltage gain A_v and dc quiescent output V_o?

12-5. If in the circuit of Fig. 12-7, $-V_{CC} = -12$ V, $R_1 = 3.3$ kΩ, $R_2 = 270$ kΩ and $R_F = 150$ kΩ, what are its gain A_v and dc quiescent output V_o? Note that the left side of R_F is at about -12 V with respect to ground.

12-6. If in the circuit of Fig. 12-7, $-V_{CC} = -9$ V, $R_1 = 2.2$ kΩ, $R_2 = 270$ kΩ and $R_F = 150$ kΩ, what are its voltage gain A_v and dc quiescent output V_o?

12-7. Referring to the circuit of Fig. 12-8, select its resistors for a gain $A_{v(dB)} = 40$ dB and a quiescent dc output $V_o = 12$ V. The supply $V_{CC} = 24$ V and the mirror current is to be 10 μA.

12-8. Show how to modify the circuit of Fig. 12-8 so that it gives a noninverting gain while working off a single negative dc supply.

12-9. What are the values of gain A_v, dc quiescent output V_o and the mirror current of the circuit of Fig. 12-9 if $V_{CC} = 20$ V, $R_1 = R_a = 4.7$ kΩ, and $R_b = R_c = R_F = 470$ kΩ?

12-10. Refer to the circuit described in the previous problem. If 200 mV of 60 Hz noise is induced into each input, as measured to the left of each capacitor with respect to ground, and 2 mV of 60 Hz noise is at the output, what is this CD amplifier's CMRR(dB)?

12-11. In the circuit of Fig. 12-13, if $V_{CC} = 12$ V, $R_1 = R_3 = 1.1$ kΩ and $R_2 = 2.2$ kΩ, what are the threshold voltages V_1 and V_2? What happens to the LED with each of the following? (a) $V_s = 0$ V, (b) $V_s = 2.5$ V, (c) $V_s = 3.5$ V, (d) $V_s = 6$ V, (e) $V_s = 8.5$ V, (f) $V_s = 9.5$ V and (g) $V_s = 12$ V.

12-12. Referring to the previous problem, if $R_1 = R_2 = R_3$, what are the values of V_1 and V_2? What happens to the LED with each of the inputs (a) through (g)?

12-13. Referring to the circuit of Fig. 12-14, if $V_{CC} = 24$ and $V_{ref} = 3$, what are the upper and lower threshold voltages and the noise margin?

12-14. Referring to the circuit of Fig. 12-14, if $V_{CC} = 15$ $V_{ref} = 5$V, and we replace R_1 with 330 kΩ, what are the upper and lower threshold voltages and the noise margin?

GLOSSARY

Amplification, Differential (A_d) The ratio of the output voltage to the differential input voltage of a differential amplifier.

Amplification, Voltage (A_v) The closed-loop voltage gain of an Op Amp which is its effective gain with negative feedback; also called *the closed-loop gain*.

Amplification, Voltage Open Loop (A_{VOL}) The ratio of the output voltage to the differential input voltage with no feedback also called *the open-loop gain*.

Average Temperature Coefficient of Input Offset Current The operating temperature range divided into the resulting change in the input offset current.

Average Temperature Coefficient of Input Offset Voltage The operating temperature range divided into the resulting change in the input offset voltage.

Bandwidth (BW) The frequency range in which the Op Amp's gain does not drop more than 0.707 or 3 dB of its dc value.

Beta (β) The ratio of the bipolar transistor's collector current to its base current; also called h_{fe}.

Bootstrapping Technique of using feedback to raise an amplifier's input impedance.

Buffering Use of an isolating circuit that prevents variations in load resistance from affecting the source of signal driving it.

Channel Separation The ratio of the output signal voltage of a driven amplifier to the output signal voltage of an adjacent undriven amplifier; usually expressed in decibels.

Chip   A small piece of semiconductor in or on which an integrated circuit is built.

Chopper-Stabilized Op Amp   A high-performance Op Amp having a very small input bias current and low drift.

Common-Mode Gain   The ratio of the common-mode output voltage to the common-mode input voltage; typically, the common-mode gain is much less than unity for differential and operational amplifiers.

Common-Mode Input Resistance   The resistance with respect to ground or a common point looking into both inputs tied together.

Common-Mode Input Voltage Swing   The peak value of common-mode input voltage that can be applied and still maintain linear operation.

Common-Mode Output Voltage   The output voltage resulting from a voltage applied to both inputs simultaneously.

Common-Mode Rejection Ratio (*CMRR*)   The ratio of the closed-loop gain to the common-mode gain; also, the ratio of the change in common-mode input voltage to the resulting change in input offset voltage; usually expressed in decibels.

Compensation   Use of externally wired components to stabilize Op Amps that are not internally compensated.

Darlington Pair   Two low-leakage bipolar transistors wired so that their total beta is the product of their individual betas.

Differential Input Resistance   Resistance seen looking into the Op Amp's inputs under open-loop condition.

Differential Gain (A_d)   See *Amplification.*

Drift   Changes in parameters with changes in temperature, supply voltage, or time.

Drive   Application of signal to the input of an amplifier.

Effective Input Resistance (R_i')   Small-signal ac input resistance seen looking into the appropriate input with the other input grounded or common under closed-loop condition.

Effective Output Resistance (R_o')   Small-signal ac resistance seen looking back into the Op Amp's output terminal under closed-loop condition.

Feedback   Use of circuitry that applies a portion of the Op Amp's output signal back to its inverting input.

FET   Field Effect Transistor.

Gain-Bandwidth Product   The product of a compensated Op Amp's closed-loop gain and its bandwidth with that gain.

Hybrid Op Amp   Op Amp containing ICs and discrete components.

IGFET   Insulated Gate Field Effect Transistor; also called a *MOSFET.*

Input Bias Current (I_B)   The average of the two dc input bias currents measured while both inputs are grounded or connected to a common point.

Input Bias Current Drift   The change in input bias current with change in temperature, supply voltage, or time.

Input Offset Current (I_{io})   The difference in the two dc input bias currents measured while both inputs are grounded or connected to a common point.

Input Offset Voltage (V_{io})   The voltage that must be applied across the inputs to force the output to zero volts under open-loop condition.

Input Resistance (R_i)   The resistance seen looking into either input with the other grounded or common under open-loop condition.

Input Voltage Range   The range of voltage that can be applied to either input over which the Op Amp is linear.

Internal Power Dissipation   The power required to operate the amplifier with an open load and no input signal.

Large-Signal Voltage Gain   The open-loop gain A_{VOL} measured while driving the Op Amp so that its output voltage swing is relatively large but not clipped.

Latch-Up   A condition where an Op Amp's output stays in saturation after the input voltage causing it is gone.

Loop Gain   The ratio of the open-loop gain to the closed-loop gain.

Monolithic ICs   An integrated circuit whose components are formed on or within a single piece of semiconductor called a substrate.

MOSFET   Metal-Oxide Semiconductor Field Effect Transistor; see *IGFET*.

Noise Figure (*NF*)   The ratio of an Op Amp's input signal-to-noise ratio to its output signal-to-noise ratio expressed in decibels.

Open-Loop Gain (A_{VOL})   The ratio of the output voltage to the differential input voltage with no feedback.

Output Offset Voltage (V_{oo})   The dc output of an Op Amp before its input is nulled; a textbook term used here for pedagogical purposes.

Output Protection   Use of a resistor in series with the output terminal to prevent excessive output currents.

Output Resistance (R_o)   Resistance seen looking back into the Op Amp's output terminal under open-loop condition.

Output Short-Circuit Current   The output current with output shorted to ground or common or connected to either dc supply.

Output Voltage Swing   The maximum output voltage swing that can be obtained without clipping.

Overshoot   The output voltage swing beyond its final quiescent value in response to a step input voltage.

Power Supply Current   The current on Op Amp draws from the power supply with the load open and no signal applied.

Power Supply Rejection Ratio (*PSRR*)   The ratio of the change in the power supply voltage to the resulting change in input offset voltage; usually expressed in decibels.

Power Supply Sensitivity   The ratio of the change in the input offset voltage to the change in power supply voltage causing it. The ratio of the change in a specified parameter to the change in the supply voltage causing it.

Rate of Closure   The rate at which the open-loop gain vs. frequency curve is decreasing (rolling off) as it intersects the closed-loop vs. frequency curve; usually expressed in decibels per octave or decibels per decade.

Ringing   The oscillations of the output voltage about the final quiescent value in response to a step input voltage.

Rise Time   The time required for an output voltage step to change from 10% to 90% of its final value.

Roll-Off   The decrease in amplifier gain at higher frequencies.

Saturation   An Op Amp driven to its maximum positive or its maximum negative output voltage; largely determined by the power supply voltages and load resistance.

Settling Time   The time between the instant a step input voltage is applied to the instant the output settles to a specified percentage of the final quiescent value.

Slew Rate   The maximum rate of change of the output voltage under large-signal conditions.

Standby Current Drain   The part of the output current of a power supply that does not contribute to the load current.

Supply Current   The current drained from the power supply to operate the amplifier with no load and with the output voltage halfway between the supplies.

Tailored Response   Term used to describe the response of Op Amps made to be externally compensated and meet special requirements not available with internally compensated types.

Temperature Stability   The change in output voltage or voltage gain over a specified temperature range.

Transfer Function   The output voltage vs. input voltage curve.

Transient Response   The output voltage response to a step input voltage under closed-loop and small-signal conditions.

Unity Gain Bandwidth   The bandwidth when the Op Amp's closed-loop gain is down to unity.

Zener Knee Current   The minimum zener current required to keep the zener in zener (avalanche) conduction.

Zener Test Current   The zener current with which the specified zener voltage and zener resistance were measured.

APPENDIX A

FIXED VOLTAGE REGULATORS

Product Type No.	Input Voltage (V) Min	Input Voltage (V) Max	Output Voltage (V) Typ	Load Regulation (mV) Typ	Line Regulation (mV) Typ	Ripple Rejection (dB) Typ	Long Term Stability (mV) Max	Output Noise Voltage (µV) Typ	Quiescent Current (mA)	Operating Temperature Range (°C) Min	Operating Temperature Range (°C) Max	Output Current† (Amps)	Package Type
Positive Regulators													
LM109K*	7	35	5	50	4	75	10	40	6	-55	125	>1	TO-3, TO-5
LM209K*	7	35	5	50	4	75	10	40	6	-25	85	>1	TO-3, TO-5
LM309K*	7	35	5	50	4	75	20	40	6	0	70	>1	TO-3, TO-5
LM123K	7.5	20	5	50	4	75	10	40	6	-55	125	3	TO-3
LM223K	7.5	20	5	50	4	75	20	40	6	-25	85	3	TO-3
LM323K	7.5	20	5	50	4	75	20	40	6	0	70	3	TO-3
LM340-05	7	35	5	100 max	100 max	70	20	40	6	0	70	>1	TO-3, TO-220
LM340-06	8	35	6	120 max	120 max	65	24	45	6	0	70	>1	TO-3, TO-220
LM340-08	10	35	8	160 max	160 max	62	32	52	6	0	70	>1	TO-3, TO-220
LM340-12	14	35	12	240 max	240 max	61	48	75	6	0	70	>1	TO-3, TO-220
LM340-15	17	35	15	300 max	300 max	60	60	90	6	0	70	>1	TO-3, TO-220
LM340-18	20	35	18	360 max	360 max	59	72	110	6	0	70	>1	TO-3, TO-220
LM340-24	26	40	24	480 max	480 max	56	96	170	6	0	70	.8	TO-3, TO-220
Negative Regulators													
LM120K-5*	-6	-25	-5	50	10	67	50	150	2	-55	125	>1	TO-3, TO-5
LM220K-5*	-6	-25	-5	50	10	67	50	150	2	-25	85	>1	TO-3, TO-5
LM320K-5*	-6	-25	-5	50	10	67	50	150	2	0	70	>1	TO-3, TO-5
LM120K-5.2*	-6.2	-25	-5.2	50	10	67	50	150	2	-55	125	>1	TO-3, TO-5
LM220K-5.2*	-6.2	-25	-5.2	50	10	67	50	150	2	-25	85	>1	TO-3, TO-5
LM320K-5.2*	-6.2	-25	-5.2	50	10	67	50	150	2	0	70	>1	TO-3, TO-5
LM120K-12*	-13	-30	-12	30	4	80	120	400	4	-55	125	>1	TO-3, TO-5
LM220K-12*	-13	-30	-12	30	4	80	120	400	4	-25	85	>1	TO-3, TO-5
LM320K-12*	-13	-30	-12	30	4	80	120	400	4	0	70	>1	TO-3, TO-5
LM120K-15*	-16	-30	-15	30	5	80	150	400	4	-55	125	>1	TO-3, TO-5
LM220K-15*	-16	-30	-15	30	5	80	150	400	4	-25	85	>1	TO-3, TO-5
LM320K-15*	-16	-30	-15	30	5	80	150	400	4	0	70	>1	TO-3, TO-5

*Ratings are for TO-3(K) package; device also available in TO-5(H) package.

†Max output current depends on package type, heat sinking, and input voltage differential.

Courtesy of National Semiconductor Corporation

Industry Package Cross-Reference Guide

	NSC	Signetics	Fairchild	Motorola	TI	RCA	Silicon General	AMD	Raytheon
14/16 Lead Glass/Metal DIP	D	I	D	L		D	D	D	D, M
Glass/Metal Flat-Pack	F	Q	F	F	F, S	K	F	F	J, F, Q
TO-99, TO-100, TO-5	H	T, K, L, DB	H	G	L	S*, V1**	T	H	T, H
8, 14 and 16-Lead Low-Temperature Ceramic DIP	J	F	R, D	L	J				DC, DD
TO-3 (Steel)	K						K		K
TO-3 (Aluminum)	KC	DA	K	K	K		K		LK, TK
8, 14 and 16-Lead Plastic DIP	N	V, A, B	T, P	P	P, N	E	M, N	PC	N, DN, DP, MP

		NSC	Signetics	Fairchild	Motorola	TI	RCA	Silicon General	AMD	Raytheon
	TO-202 (D-40, Durawatt) **(Package 37)**	P								
	"SGS" Type Power DIP **(Package 39)**	S		BP						
	TO-220 **(Package 26)**	T	U	U		KC				
	Low Temperature Glass Hermetic Flat Pack	W		F	F	W			FM	
	TO-92 (Plastic) **(Package 38)**	Z	S	W	P	LP				

*With dual-in-line formed leads.

**With radially formed leads.

Courtesy of National Semiconductor Corporation

APPENDIX C

Ordering Information

PACKAGE

D — Glass/Metal Dual-In-Line Package

F — Glass/Metal Flat Pack

H — TO-5 (TO-99, TO-100, TO-46)

J — Low Temperature Glass Dual-In-Line Package

K — TO-3 (Steel)

KC — TO-3 (Aluminum)

N — Plastic Dual-In-Line Package

P — TO-202 (D-40, Durawatt)

S — "SGS" Type Power Dual-In-Line Package

T — TO-220

W — Low Temperature Glass Flat-Pack

Z — TO-92

DEVICE NUMBER

3, 4, or 5 Digit Number Suffix Indicators:

A — Improved Electrical Specification

C — Commercial Temperature Range

```
LM   101A   F
              └PACKAGE

            DEVICE NUMBER

            DEVICE FAMILY
```

DEVICE FAMILY

AD — Analog to Digital

AH — Analog Hybrid

AM — Analog Monolithic

CD — CMOS Digital

DA — Digital to Analog

DM — Digital Monolithic

LF — Linear FET

LH — Linear Hybrid

LM — Linear Monolithic

LX — Transducer

MM — MOS Monolithic

TBA — Linear Monolithic

Devices are listed in the table of contents alpha-numerically by device family (LH, LM, LX, etc.) and then by device number. **With most of National's proprietary linear circuits, a 1-2-3 numbering system is employed. The 1 denotes a Military temperature range device (−55°C to +125°C), the 2 denotes an Industrial temperature range device (−25°C to +85°C), and the 3 denotes a Commercial temperature range device (0°C to +70°C), i.e. LM101/LM201/LM301.**

Exceptions to this are the LM1800 series of consumer circuits which are specified for the commercial temperature range; some hybrid circuits which employ a "C" suffix to denote the commercial temperature range; and second-source products which follow the original manufacturers numbering system, i.e. LM741/LM741C or LM1414/LM1514.

Parts are generally listed in the table of contents by military part number first, i.e. LM139/LM239/LM339. Where a separate data sheet exists for a different temperature range, the device will be listed separately, i.e. LM119/LM219 and listed separately LM319. Where only one temperature range exists, the part will be listed in its proper order, i.e. LM340.

Courtesy of National Semiconductor Corporation

APPENDIX D

Device	Input Offset Voltage Max (mV)	Input Offset Voltage Drift Max (μV/°C)	Input Offset Current Max (nA)	Input Bias Current Max (nA)	Voltage Gain Min (Volts/V)	Bandwidth Av=1 Typ (MHz)	Slew Rate Av=1 Typ (V/μs)	Output Voltage Swing R_L=10kΩ (V)	Supply Voltage Min (V)	Supply Voltage Max (V)	Common Mode Rejection Ratio (dB) Min #	Differential Input Voltage (V)	Supply Current TA=25°C Max (mA) (Note 2)	Compensation Components	Package Types
SINGLE OP AMPS															
LM201	10	10 typ	750	200	15k	1	0.5	5	±3	±22	±12	±30	3	1	TO-5 F. P.
LM301A	10	30	70	300	15k	1	0.5	5	±3	±18	±12	±30	3	1	TO-5 DIP
LM302	20	20 typ	*	30	0.9985	10	10	1	±12	±18	±10	*	5.5	0	TO-5
LM307	10	30	50	250	15k	1	0.5	5	±3	±18	±12	±30	3	0	TO-5 DIP F.P.
LM308A	0.73	5	1.5	10	60k	1	0.3	1	±2	±20	±14	(Note 1)	0.8	1	TO-5 DIP F.P.
LM308	10	30	1.5	10	15k	1	0.3	1	±2	±18	±14	(Note 1)	0.8	1	TO-5 DIP F.P.
LM310	10	10 typ	*	10	0.999	20	30	1	±5	±18	±10	*	5.5	0	TO-5 DIP F.P.
LM312	10	30	1.5	10	15k	1	0.3	1	±2	±18	±14	(Note 1)	0.8	0	TO-5 DIP F.P.
LM316A	6	*	0.03	0.1	30k	1	0.3	1	±5	±20	±13	(Note 1)	0.6	0	TO-5 DIP F.P.
LM316	15	*	0.1	0.25	15k	1	0.3	1	±5	±20	±13	(Note 1)	0.8	0	TO-5 DIP F.P.
LM318	15	*	300	750	20k	15	50	5	±5	±18	±11.5	(Note 1)	10	0	TO-5 DIP
LM321A	0.65	0.2	1	25	12k	0.5	*	*	±5	±20	±15	±15	2.2	1	TO-5 DIP F.P.
(RSET = 70k)															
LM321	2.5	1	4	28	12k	0.5	*	*	±5	±20	±15	±15	2.2	1	TO-5 DIP F.P.
(RSET = 70k)															
LM343	10	*	14	55	50k	1	2.5	4 (RL ≥ 5k)	±4	±34	±34	±34	5.0	0	TO-5 DIP F.P.
LM344	10	*	14	55	50k	2	30	4 (RL ≥ 5k)	±4	±34	±34	±34	5.0	1	TO-5 DIP F.P.
LF351	10	10 typ	0.1	0.2	25k	4	13	±12	-18	18	70	±30	3.4	0	H, N
LF351A	2	10 typ	0.05	0.2	50k	4	13	±12	-18	18	80	±30	2.8	0	H, N
LF351B	5	10 typ	0.1	0.1	50k	4	13	±12	-18	18	80	±30	2.8	0	H, N
LF355A	2.3	5	1	5	25k	2.5	5	5	±5	±22	±20	±40	4	0	TO-5, Mini-DIP
LF355	13	5 typ	2	8	15k	2.5	5	5	±5	±18	±16	±30	4	0	TO-5, Mini DIP
LF356A	2.3	5	1	5	25k	5	15	5	±5	±22	±20	±40	10	0	TO-5, Mini-DIP
LF356	13	5 typ	2	8	25k	5	15	5	±5	±18	±16	±30	10	0	TO-5, Mini-DIP
LF357A	2.3	5	1	5	25k	25	75	5	±5	±22	±20	±40	10	0	TO-5, Mini-DIP
(Av ≥ 5)															
LF357	13	5 typ	2	8	15k	25	75	5	±5	±18	±16	±30	10	0	TO-5, Mini DIP
(Av ≥ 5)															
LF13741	20	10 typ	2	8	15k	1	0.5	5	±4	±18	±16	±30	4	0	TO-5, Mini-DIP

Note 1: Inputs have shunt-diode protection; current must must be limited. *Not specified

This column shows either the maximum common-mode input voltage or the minimum CMRR.

Courtesy of National Semiconductor Corporation

251

COMMERCIAL TEMPERATURE RANGE 0°C ≤ TA ≤ +70°C

Device	Input Offset Voltage Max (mV)	Input Offset Voltage Drift Max (μV/°C)	Input Offset Current Max (nA)	Input Bias Current Max (nA)	Voltage Gain Min (Volts/V)	Bandwidth Av=1 Typ (MHz)	Slew Rate Av=1 Typ (V/μs)	Output Voltage Swing RL=10 kΩ (V)	Supply Voltage Min (V)	Supply Voltage Max (V)	Common Mode Rejection Ratio (dB) Min #	Differential Input Voltage (V)	Supply Current TA=25°C Max (mA) (Note 2)	Compensation Components	Package Types
SINGLE OP AMPS (Continued)															
LM709C	10	12 typ	500	1500	15k	1	0.3	5	±9	±18	±8	±5	6.6	3	TO-5 DIP
LM725C	3.5	2 typ	50	250	125k	0.5	0.005	5	±3	±22	±13.5	±5	5	4	TO-5 DIP
LM741C	7.5	15 typ	300	800	15k	0.5	0.5	5	±3	±18	±12	±30	2.8	0	TO-5 DIP F.P.
LM741E	4	15	70	210	32k	1	0.5	7.5	±3	±18	±12	±30	3.75	0	TO-5 DIP F.P.
LM748C	6	6	0.5	1.5	25k	1	0.5	5	±3	±18	±12	±30	2.8	1	TO-5 DIP
LM4250C	6	*	8	10	50k	0.1	0.03	0.12	±1	±18	±12	±15	0.011 (Set)	0	TO-5 DIP
							(Av >10)	(RL ≥100k)							
DUAL OP AMPS															
LF353	10	10 typ	0.1	0.2	25k	4	13	±12	-18	18	70	±30	3.4	0	N, H
LF353A	2	10 typ	0.05	0.2	50k	4	13	±12	-18	18	80	±30	2.8	0	N, H
LF353B	5	10 typ	0.1	0.1	50k	4	13	±12	-18	18	80	±30	2.8	0	N H
LM358	7.5	7 typ	150	500	15k	1	*	8	±1.5	±15	V⁺-1.5	V⁺	1.2	0	TO-5 DIP
LM1458	6	*	300	800	15k	1	0.2	5	±3	±18	±15	±30	5.6	0	TO-5 DIP
LM747C	6	*	300	800	15k	1	0.5	5	±3	±18	±12	±30	5.6	0	DIP
LM747E	4	15	70	210	32k	1	0.5	7.5	±3	±18	±12	±30	5.6	0	DIP
QUAD OP AMPS															
LF347	10	10 typ	0.01	0.2	25k	4	13	±12	-18	18	70	±30	3.4	0	N, J
LF347A	2	10 typ	0.05	0.2	50k	4	13	±12	-18	18⁻	80	±30	2.8	0	N, J
LF347B	5	10 typ	0.1	0.1	50k	4	13	±12	-18	18	80	±30	2.8	0	N, J
LM324	9	7 typ	150	500	15k	1	*	10-source 5-sink	3 (±1.5)(±16)	32	V⁺-1.5	32	2	0	DIP F.P.
LM346	5	10 typ	100	250	100k	0.8	0.4	±12	-18	18	70	±30	0.62	0	N, J
LM348	7.5	15 typ	100	400	15k	1	*	5	±5	±18	±18	±36	4.5	0	DIP F.P.
LM349 (Av ≥ 5)	7.5	15 typ	100	400	15k	4	3	5	±5	±18	±18	±36	4.5	0	DIP F.P.
LM3900	*	*	*	200	2.8k	2.5	20	10	4 (±2)	36 (±18)	*	*	10	0	DIP

Note 2: Supply current for all channels of amplifier in the package

This column shows either the maximum common-mode input voltage or the minimum CMRR.

SELECTION GUIDE
FOR GENERAL-PURPOSE
OPERATIONAL AMPLIFIERS

Selection Guide for Commercial Operational Amplifiers

General Purpose — Tailored Response

		μA702*	μA709	μA739	μA748	μA749	μA777	201	201A	301A
		DC Wide Band		Dual Low Noise	Dual					
Input Offset Voltage	Max (mV)	5.0	7.5	6.0	6.0	6.0	7.5	7.5	2.0	7.5
Input Offset Current	Max (nA)	2000	500	1000	200	500	50	200	10	50
Input Bias Current	Max (nA)	7500	1500	2000	500	1000	250	500	75	250
Voltage Gain	Min (V/mV)	2.0	15	6.5	20	15	25	20	25	25
Operating Supply Voltage Range	Min (V)	+6.0, −3.0	±9.0	±4.0	±5.0	±4.0	±5.0	±5.0	±5.0	±5.0
	Max (V)	+14, −7.0	±18	±18	±18	±18	±20	±20	±20	±20
Unity Gain Bandwidth	Typ (MHz)	30	1.0	10	1.0	10	1.0	1.0	1.0	1.0
Slew Rate $A_{CL} = 1$	Typ (V/μs)	3.5	0.3	1.0	0.5	1.5	0.5	0.5	0.5	0.5
$A_{CL} = -1$		3.5	0.3	2.5	6.0	2.5	6.0	6.0	6.0	6.0
$A_{CL} = 10$		5.0	3.0	8.0	2.0	8.0	2.0	2.0	2.0	2.0
Input Voltage Range	Max (V)	+1.5, −6.0	±10	±15	±15	±15	±15	±15	±15	±15
Differential Input Voltage	Max (V)	±5.0	±5.0	±5.0	±30	±5.0	±30	±30	±30	±30
Input Offset Voltage Drift	Typ (μV/°C)	10	10	4.0	7.0	3.0	3.0	7.0	3.0	6.0
Offset Adjust					X		X	X	X	X
Output Short Circuit Protection					X		X	X	X	X
Compensated										
Dual				X		X				

Selection Guide for Military Operational Amplifiers

General Purpose — Tailored Response

		μ702*	μA709A	μA709	μA748	μA749	μA777	101	101A
		DC Wide Band			Dual				
Input Offset Voltage	Max (mV)	2.0	2.0	5.0	5.0	3.0	2.0	5.0	2.0
Input Offset Current	Max (nA)	500	50	200	200	400	10	200	10
Input Bias Current	Max (nA)	5000	200	500	500	750	75	500	75
Voltage Gain	Min (V/mV)	2.5	25	25	50	25	50	50	50
Operating Supply Voltage Range	Min (V)	+6.0, −3.0	±9.0	±9.0	±5.0	±4.0	±5.0	±5.0	±5.0
	Max (V)	+14, −7.0	±18	±18	±22	±18	±20	±20	±20
Unity Gain Bandwidth	Typ (MHz)	30	5.0	5.0	1.0	10	1.0	1.0	1.0
Slew Rate $A_{CL} = 1$	Typ (V/μs)	3.5	0.3	0.3	0.5	1.5	0.5	0.5	0.5
$A_{CL} = -1$		3.5	0.3	0.3	6.0	2.5	6.0	6.0	6.0
$A_{CL} = 10$		5.0	3.0	3.0	2.0	8.0	2.0	2.0	2.0
Input Voltage Range	Max (V)	+1.5, −6.0	±10	±10	±15	±15	±15	±15	±15
Differential Input Voltage	Max (V)	±5.0	±5.0	±5.0	±30	±5.0	±30	±30	±30
Input Offset Voltage Drift	Typ (μV/°C)	10	1.8	3.0	7.0	3.0	3.0	3.0	3.0
	Max (μV/°C)						15		15
Offset Adjust					X		X	X	X
Output Short Circuit Protection					X		X	X	X
Compensated									
Dual						X			

* $V_S = +12, -6.0$ V
$V_S = \pm 15$ V, $T_A = 25°C$ unless otherwise specified

			General Purpose					High Power
307	310	µA741	Compensated µA741E	µA776	µA747	1458		µA791
	Voltage Follower	Industry Standard	High Performance	Programmable $I_{SET} = 15\,\mu A$	Dual Industry Standard	Dual		1 Amp
7.5	7.5	6.0	3.0	6.0	6.0 \	6.0		6.0
50	—	200	30	25	200	200		200
250	7.0	500	80	50	500	500		500
25	$.999 \times 10^{-3}$	20	50	50	20	20		20
±5.0	±5.0	±5.0	±5.0	±1.2	±5.0	±5.0		±5.0
±18	±18	±18	±22	±18	±18	±18		±18
1.0	20	1.0	1.0	1.0	1.0	1.0		0.2
0.5	30	0.5	0.7	0.7	0.5	0.5		0.5
0.5	—	0.5	0.7	0.7	0.5	0.5		1.0
0.5	—	0.5	0.7	0.7	0.5	0.5		6.0
±15	±10	±15	±15	±15	±15	±15		±15
±30	—	±30	±30	±30	±30	±30		±30
6.0	10	7.0	3.0	3.0	7.0	7.0		15
X	X	X	X	X	X			X
X	X	X	X	X	X	X		X
X	X	X	X	X	X	X		
					X	X		

			General Purpose					High Power
107	110	µA741	Compensated µA741A	µA776	µA747	1558		µA791
	Voltage Follower	Industry Standard	High Performance	Programmable $I_{SET} = 15\,\mu A$	Dual Industry Standard	Dual		1 Amp
2.0	4.0	5.0	3.0	5.0	5.0	5.0		5.0
10	—	200	30	15	200	200		200
75	3.0	500	80	50	500	500		500
50	$.999 \times 10^{-3}$	50	50	50	50	50		50
±5.0	±5.0	±5.0	±5.0	±1.2	±5.0	±5.0		±5.0
±20	±18	±22	±22	±18	±22	±22		±22
1.0	20	1.0	1.0	1.0	1.0	1.0		0.2
0.5	30	0.5	0.6	0.7	0.5	0.5		0.5
0.5	—	0.5	0.6	0.7	0.5	0.5		1.0
0.5	—	0.5	0.6	0.7	0.5	0.5		6.0
±15	±10	±15	±15	±15	±15	±15		±15
±30	—	±30	±30	±30	±30	±30		±30
3.0	6.0	7.0	5.0	3.0	7.0	7.0		10
15			15					
	X	X	X	X	X			X
X	X	X	X	X	X	X		X
X	X	X	X	X	X	X		
					X	X		

Courtesy of Fairchild Semiconductor

COLLECTION OF
OP AMP TYPES
AND THEIR PARAMETERS

LINEAR OPERATIONAL AMPLIFIERS
$T_A = 25°C$

Device	Operating Temp. Range °C Min	Max	A_{VOL} Min K	R_i Min Ω	P_D mW	$|I_{io}|$ Max nA	I_b Max nA	CMV_i Min V	†Typ. Slew Rate SR V/μSec	V_o Min V	$|V_{io}|$ Max mV	Offset Adjust	Internal Compensation	Output Protection	Input Protection	JEDEC Package Type
709A	−55	+125	25	350K	108	50	200	±8	0.3	±12	1	no	no	no	no	TO-91, 99, 116
709B	−55	+125	25	150K	165	200	500	±8	0.3	±12	5	no	no	no	no	TO-91, 99, 116
709C	0	+70	15	50K	200	500	1500	±8	0.3	±12	10	no	no	no	no	TO-91, 99, 116
739C(Dual)	0	+70	6.5	37K	420	1000	2000	±10	1.0	+12, −14	6	no	no	yes	yes	TO-116
741B	−55	+125	50	300K	85	200	500	±12	0.5	±12	5	yes	yes	yes	yes	TO-91, 99, 116
741C	0	+70	20	150K	85	200	500	±12	0.5	±12	6	yes	yes	yes	yes	TO-91, 99, 116
747B(Dual)	−55	+125	50	300K	85	200	500	±12	0.5	±12	5	yes	yes	yes	yes	TO-101, 116
747C(Dual)	0	+70	20	150K	85	200	500	±12	0.5	±12	6	yes	yes	yes	yes	TO-101, 116
748B	−55	+125	50	300K	85	200	500	±12	0.5	±12	5	yes	no	yes	yes	TO-99
748C	0	+70	20	150K	85	200	500	±12	0.5	±12	6	yes	no	yes	yes	TO-99
749B(Dual)	−55	+125	25	100K	220	400	750	±11	1.5	+12, −14.5	3	no	no	yes	yes	TO-116
749C(Dual)	0	+70	15	70K	330	500	1000	±11	1.5	+12, −14.5	6	no	no	yes	yes	TO-116
800B	−55	+125	10	250K	180	100	1000	±4	—	±6	50	no	no	no	no	TO-101
800D	−55	+125	10	100K	180	200	2000	±4	—	±12	—	no	no	no	no	TO-101
801B	−55	+125	10	250K	180	100	1000	±4	—	±12	50	no	no	no	no	TO-100
801D	−55	+125	10	100K	180	200	2000	±4	—	±12	—	no	no	no	no	TO-100
805B	−55	+125	30	500K	225	50	500	±8	2.5	±12	5	no	no	no	yes	TO-91, 99
805C	0	+100	10	100	225	100	1000	±8	2.5	±12	10	no	no	no	yes	TO-91, 99
806B	−55	+125	30	500	225	50	500	±8	2.5	±9	5	no	no	no	yes	TO-91, 99
806C	0	+100	10	100	225	100	1000	±8	2.5	±9	10	no	no	no	yes	TO-91, 99
807B	−55	+125	30	500	225	50	500	±8	2.5	±12	2.5	no	no	no	yes	TO-91, 99
808A	−55	+125	25	1M	225	15	50	±8	2.5	±12	5	no	no	no	yes	TO-91, 99
808B	−55	+125	25	1M	225	30	50	±8	2.5	±12	10	no	no	no	yes	TO-91, 99
809B	−55	+125	10	100	150	100	500	±10	—	±10	10	no	no	yes	yes	TO-99, 116
809C	0	+100	10	50	150	350	1000	±10	—	±10	10	no	no	yes	yes	TO-99, 116
810B(Dual)	−55	+125	10	100	150	100	500	±10	—	±10	10	no	no	yes	yes	TO-116
810C(Dual)	0	+100	10	50	150	350	1000	±10	—	±10	10	no	no	yes	yes	TO-116

†Unity Gain

Courtesy of Teledyne Semiconductor

LINEAR OPERATIONAL AMPLIFIERS
$T_A = 25°C$ (Cont.)

Device	Operating Temp. Range °C Min	Max	A_{VOL} Min K	R_i Min Ω	P_D mW	$|I_{io}|$ Max nA	I_b Max nA	CMV_i Min V	†Typ. Slew Rate SR V/μSec	V_o Min V	$|V_{io}|$ Max mV	Offset Adjust	Internal Compensation	Output Protection	Input Protection	JEDEC Package Type
715B	−55	+125°C	15	IM(typ)	210	250	750	±15	18	±10	±5	yes	yes	yes	yes	TO-100
715C	0	+70°C	10	IM(typ)	300	1500	250	±15	18	±10	±7.5	yes	no	yes	yes	TO-100
846B	−55	+125	100	25 M	90	5	30	±12.5	2.0	±12	±3	yes	no	yes	yes	TO-99
846C	0	+70	50	15 M	75	15	50	±12	2.0	±12	±5	yes	no	yes	yes	TO-99
LM101A	−55	+125	50	1.5 M	120	10	25	±12	0.5	±12	±2	yes	no	yes	yes	TO-99
LM101B	−55	+125	50	300 K	120	200	500	±12	0.5	±12	±5	yes	no	yes	yes	TO-99
LM201A	−25	+85	50	1.5 M	120	10	75	±12	0.5	±12	±2	yes	no	yes	yes	TO-99
LM201C	−25	+85	20	300 K	90	200	500	±12	0.5	±12	±7.5	yes	no	yes	yes	TO-99
LM301A	0	+70	25	500 K	120	50	250	±12	0.5	±12	±7.5	yes	no	no	yes	TO-99
LM307D	0	+70	25	500 K	120	50	250	±12	0.5	±12	±7.5	yes	no	no	yes	TO-99
811B	−55	+125	10	100	150	100	500	±10	—	±10	10	no	no	yes	yes	TO-99, 116
811C	0	+100	10	50	—	350	1000	±10	—	±10	10	no	no	yes	yes	TO-99, 116
813C	0	+70	6	—	120	2000	5000	±5	—	—	4	no	no	yes	yes	TO-99, 116
819B	−55	+125	5	50K	25	100	500	±4		±4	10	no	no	yes	yes	TO-99
841B	−55	+125	50	300	85	200	500	±12	0.5	±12	5	no	no	yes	yes	TO-99
841C	0	+100	20	150K	85	200	500	±12	0.5	±12	6	no	no	yes	yes	TO-99
844B	−55	+125	100	25 M	75	5	30	±12.5	2.0	±12	±3	yes	yes	yes	yes	TO-99
844C	0	+70	50	15 M	90	15	50	±12	2.0	±12	±5	yes	yes	yes	yes	TO-99
LM107B	−55	+125	50	1.5 M	120	10	75	±12	0.5	±12	±2	yes	yes	yes	yes	TO-99
LM207C	−25	+85	50	1.5 M	120	10	75	±12	0.5	±12	±2	yes	yes	yes	yes	TO-99
MC1437C(Dual)	0	+70	15	50K	200	500	1500	±8	0.3	±12	10	no	no	no	no	TO-116
MC1439C	0	+70	15	100	200	100	1000	±11	4.2	±10	7.5	no	no	yes	yes	TO-99, 116
MC1458C(Dual)	0	+70	20	150K	85	200	500	±12	0.5	±12	6	no	yes	yes	yes	TO-99, 116
MC1537B(Dual)	−55	+125	25	150K	165	200	500	±8	0.3	±12	5	no	no	no	no	TO-91, 99, 116
MC1539B	−55	+125	50	150	150	60	500	±11	4.2	±10	3	no	no	yes	yes	TO-99, 116
MC1558B(Dual)	−55	+125	50	300K	85	200	500	±12	0.5	±12	5	yes	yes	yes	yes	TO-101, 116

†Unity Gain

APPENDIX G

*SELECTION GUIDE
FOR HIGH ACCURACY
OPERATIONAL AMPLIFIERS*

Selection Guide for Commercial Operational Amplifiers

High Accuracy Instrumentation

		FET μA740	Bipolar μA776	μA777	Super Beta 208	308	208A	308A
		Low Bias Current						
		High Z_{IN} $10^{12}\,\Omega$ High Slew Rate	Low Power $I_{SET} = 1.5\,\mu A$					
Input Offset Voltage	Max (mV)	100	6.0	7.5	2.0	7.5	0.5	0.5
Input Offset Current	Max (nA)	0.3	6.0	50	0.2	1.0	0.2	1.0
Input Bias Current	Max (nA)	2.0	10	250	2.0	7.0	2.0	7.0
Voltage Gain	Min (V/mV)	25	50	25	50	15	80	80
Operating Supply Voltage Range	Min (V)	± 5.0	± 1.2	± 5.0	± 5.0	± 5.0	± 5.0	± 5.0
	Max (V)	± 22	± 18	± 20	± 20	± 18	± 20	± 20
Unity Gain Bandwidth	Typ (MHz)	3.0	0.2	1.0	1.0	1.0	1.0	1.0
Slew Rate $A_{CL} = 1$	Typ (V/μs)	6.0	0.1	0.5	0.3	0.3	0.3	0.3
$A_{CL} = -1$		6.0	0.1	6.0	0.6	0.6	0.6	0.6
$A_{CL} = 10$		6.0	0.1	2.0	—	—	—	—
Input Voltage Range	Max (V)	± 15	± 15	± 15	± 15	± 15	± 15	± 15
Differential Input Voltage	Max (V)	± 30	± 30	± 30	± 0.5	± 0.5	± 0.5	± 0.5
Input Offset Voltage Drift	Typ (μV/°C)	20	3.0	3.0	3.0	6.0	1.0	1.0
	Max (μV/°C)	—	—	—	15	30	5.0	5.0
Offset Adjust		X	X	X				
Output Short Circuit Protection		X	X	X	X	X	X	X
Compensated		X	X					

Selection Guide for Military Operational Amplifiers

High Accuracy Instrumentation

		FET μA740	Bipolar μA776	μA777	Super Beta 108	108A
		Low Bias Current				
		High Z_{IN} $10^{12}\,\Omega$ High Slew Rate	Low Power $I_{SET} = 1.5\,\mu A$			
Input Offset Voltage	Max (mV)	20	5.0	2.0	2.0	0.5
Input Offset Current	Max (nA)	0.15	3.0	10	0.2	0.2
Input Bias Current	Max (nA)	0.2	7.5	75	2.0	2.0
Voltage Gain	Min (V/mV)	50	50	50	50	80
Operating Supply Voltage Range	Min (V)	± 5.0	± 1.2	± 5.0	± 5.0	± 5.0
	Max (V)	± 22	± 18	± 20	± 20	± 20
Unity Gain Bandwidth	Typ (MHz)	3.0	0.2	1.0	1.0	1.0
Slew Rate $A_{CL} = 1$	Typ (V/μs)	6.0	0.1	0.5	0.3	0.3
$A_{CL} = -1$		6.0	0.1	6.0	0.6	0.6
$A_{CL} = 10$		6.0	0.1	2.0	—	—
Input Voltage Range	Max (V)	± 15	± 15	± 15	± 15	± 15
Differential Input Voltage	Max (V)	± 30	± 30	± 30	± 0.5	± 0.5
Input Offset Voltage Drift	Type (μV/°C)	20	3.0	3.0	3.0	1.0
	Max (μV/°C)	—	—	15	15	5.0
Offset Adjust		X	X	X		
Output Short Circuit Protection		X	X	X	X	X
Compensated		X	X			

		High Accuracy Instrumentation							High Speed			
		Low Drift										
	Op Amps				Preamps							
μA725C	μA725E	μA741E	208A	308A	μA726	μA727	μA715	μA748	μA776	μA777	301A	310
					$I_C = 10\,\mu A$ Temp. 0 to +85°C	Temp. −20 to +85°C	Feed-Forward	Programmable	$I_{SET} = 500\,\mu A$	Feed-Forward	Feed-Forward	Voltage Follower
2.5	0.5	3.0	0.5	0.5	3.0	10	7.5	6.0	6.0	7.5	7.5	7.5
35	5.0	30	100	1.0	100	25	250	200	6.0	50	50	—
125	75	80	300	7.0	300	75	1500	500	10	250	250	7.0
250	1000	50	—	80	—	0.06	10	20	50	25	25	$.999 \times 10^{-3}$
±3.0	±3.0	±5.0	±5.0	±5.0	±5.0	±9.0	±6.0	±5.0	±1.2	±5.0	±5.0	±5.0
±22	±22	±22	±18	±20	±18	±18	±18	±18	±18	±20	±20	±18
1.0	1.0	1.0	1.0	1.0	20	1.0	65	1.0	1.2	1.0	1.0	20
—	—	0.6	0.3	0.3	—	—	18	0.5	15	0.5	0.5	30
—	—	0.6	0.6	0.6	—	—	100	6.0	15	6.0	15	—
—	—	0.6	—	—	—	—	38	2.0	15	2.0	5.0	—
±22	±22	±15	±15	±15	±30	±10	±15	±15	±15	±15	±15	±15
±22	±22	±30	±0.5	±0.5	±5.0	±15	±15	±30	±30	±30	±30	—
0.5	0.5	4.0	1.0	1.0	0.2	0.6	6.0	7.0	3.0	3.0	6.0	10
5.0	2.0	15	5.0	5.0	1.0	1.5	1.0					
X	X	X		X	X		X	X	X	X	X	X
X	X	X		X	X	X		X	X	X	X	X
		X		X	X							

		High Accuracy Instrumentation							High Speed			
		Low Drift										
	Op Amps				Preamps							
μA725	μA725A	μA741A	108A	μA726	μA727	μA715	μA748	μA776	μA777	101A	110
				$I_C = 10\,\mu A$		Feed-Forward	Programmable	$I_{SET} = 500\,\mu A$	Feed-Forward	Feed-Forward	Voltage Follower
1.0	0.5	3.0	0.5	2.5	10	5.0	5.0	5.0	2.0	2.0	4.0
20	5.0	30	0.2	50	15	250	200	3.0	10	10	—
100	75	80	2.0	150	40	750	500	7.5	75	75	3.0
1000	1000	50	80	—	0.06	15	50	50	50	50	$.999 \times 10^{-3}$
±3.0	±3.0	±5.0	—	±5.0	±9.0	±6.0	±5.0	±1.2	±5.0	±5.0	±5.0
±22	±22	±22	±20	±18	±18	±18	±22	±18	±20	±20	±18
1.0	1.0	1.0	1.0	20	1.0	65	1.0	1.2	1.0	1.0	20
—	—	0.6	0.3	—	—	18	0.5	15	0.5	0.5	30
—	—	0.6	0.6	—	—	100	6.0	15	6.0	6.0	—
—	—	0.6	—	—	—	38	2.0	15	2.0	2.0	—
±22	±22	±15	±15	±30	±10	±15	±15	±15	±15	±15	±15
±22	±22	±30	±0.5	±5.0	±15	±15	±30	±30	±30	±30	—
0.5	0.5	3.0	1.0	0.2	0.6	6.0	7.0	3.0	3.0	3.0	6.0
5.0	2.0	15	5.0	1.0	1.5					15	
X	X	X		X		X	X	X	X	X	X
X	X	X		X	X		X	X	X	X	X
		X	X	X							

Courtesy of Fairchild Semiconductor

SPECIFICATIONS
OF THE
709 OP AMP

	Parameter	Minimum	Typical	Maximum
A_{VOL}	Open loop voltage gain	15,000	45,000	
V_{io}	Input offset voltage		1 mV	7.5 mV
I_B	Input bias current		200 nA	1500 nA
I_{io}	Input offset current		50 nA	500 nA
R_o	Output resistance		150 Ω	
CMR (dB)	Common mode rejection	65 dB	90 dB	
R_i	Input resistance	50 kΩ	250 kΩ	

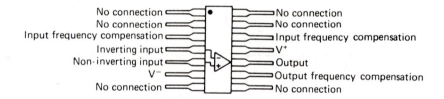

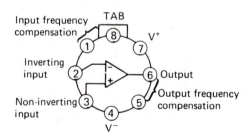

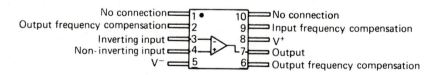

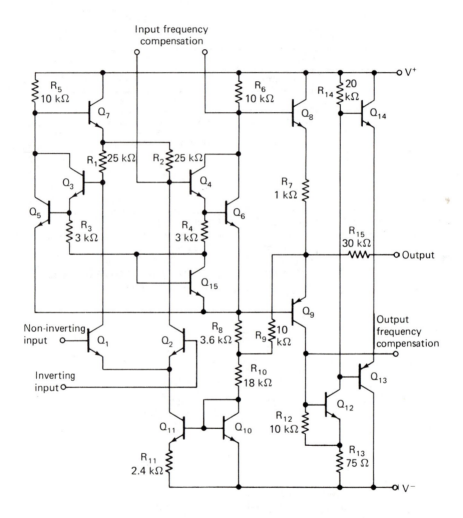

Figure 1

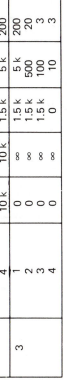

Fig. no.	Curve no.	Test conditions				
		R_1 (Ω)	R_2 (Ω)	R_3 (Ω)	C_1 (pF)	C_2 (pF)
2	1	1 k	1 M	0	10	3
	2	10 k	1 M	1.5 k	100	3
	3	10 k	100 k	1.5 k	500	20
	4	10 k	10 k	1.5 k	5 k	200
3	1	0	∞	1.5 k	5 k	200
	2	0	∞	1.5 k	500	20
	3	0	∞	1.5 k	100	3
	4	0	∞	0	10	3

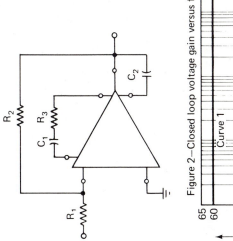

Figure 2—Closed loop voltage gain versus frequency

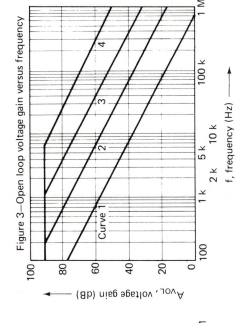

Figure 3—Open loop voltage gain versus frequency

APPENDIX I

SPECIFICATIONS
OF THE
741 OP AMP

	Parameters	Minimum	Typical	Maximum
A_{VOL}	Open loop voltage gain	50,000	200,000	
V_{io}	Input offset voltage		1 mV	6 mV
I_B	Input bias current		80 nA	500 nA
I_{io}	Input offset current		20 nA	200 nA
R_o	Output resistance		75 Ω	
CMR (dB)	Common mode rejection	70 dB	90 dB	
R_i	Input resistance	300 kΩ	2 MΩ	

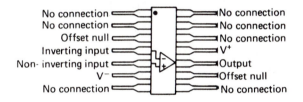

No connection — No connection
No connection — No connection
Offset null — No connection
Inverting input — V^+
Non- inverting input — Output
V^- — Offset null
No connection — No connection

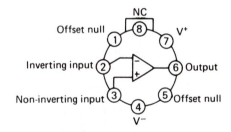

NC

Offset null — (1)(8) — V^+ (7)
Inverting input (2) — (6) Output
Non-inverting input (3) — (5) Offset null
(4)
V^-

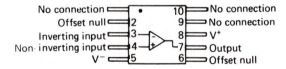

No connection — 10 — No connection
Offset null — 2 — 9 — No connection
Inverting input — 3 — 8 — V^+
Non- inverting input — 4 — 7 — Output
V^- — 5 — 6 — Offset null

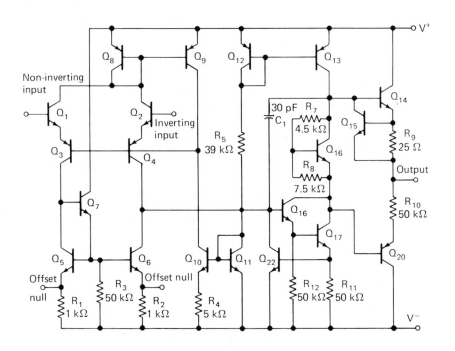

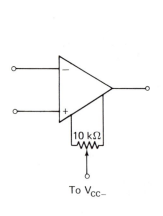

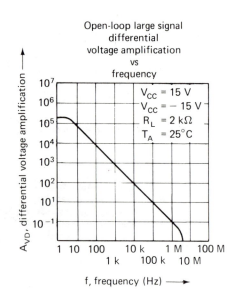

Open-loop large signal differential voltage amplification vs frequency

DERIVATION OF

EQUATION (4-4)

As shown in the circuit of Fig. J-1, the bias current I_{B_1} flows through two paths, R_1 and R_F. This current sees the resistors R_1 and R_F in parallel and causes a voltage drop across them, which is the voltage at the inverting input 1 to ground. This can be shown as

$$V_1 = \left(\frac{R_1 R_F}{R_1 + R_F} \right) I_{B_1}$$

The current I_{B_2} sees no resistance between the noninverting input 2 and ground, and therefore the voltage at this input is $V_2 = 0$ V to ground. Thus the differential input voltage caused by these currents is $V_1 - V_2 = V_1$. This voltage V_1 is amplified by the circuit's closed-loop gain, resulting in an

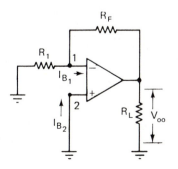

Figure J-1

output offset of

$$V_{oo} = A_v V_1 \cong -\frac{R_F}{R_1} V_1 = -\frac{R_F}{R_1}\left(\frac{R_1 R_F}{R_1 + R_F}\right) I_{B_1} = \frac{-R_F R_F}{R_1 + R_F}(I_{B_1})$$

Since R_F^2 is much larger than $R_1 + R_F$, the above simplifies to

$$V_{oo} \cong \frac{R_F^2}{R_F}(I_{B_1}) = R_F I_{B_1} \tag{4-4}$$

where V_{oo} is positive if I_{B_1} flows into the Op Amp, as it does in most bipolar input Op Amps.

DERIVATION OF
EQUATION (4-7)

As shown in the circuit of Fig. K-1, the bias current I_{B_1} flows through two paths, R_1 and R_F. This current sees essentially resistors R_1 and R_F in parallel and causes a voltage drop across them and at the input 1 with respect to ground, that is

$$V_1 = \left(\frac{R_1 R_F}{R_1 + R_F} \right) I_{B_1}$$

The current I_{B_2} flows through resistance R_2, causing input 2 to be off ground by

$$V_2 = R_2 I_{B_2} = \left(\frac{R_1 R_F}{R_1 + R_F} \right) I_{B_1}$$

The difference in these voltages, $V_1 - V_2$, is the differential input which is

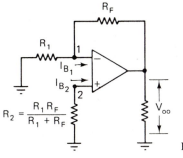

Figure K-1

amplified by the circuit's closed-loop gain—that is,

$$V_{oo} = A_v(V_1 - V_2) = -\frac{R_F}{R_1}\left[\frac{R_1 R_F}{R_1 + R_F}(I_{B_1}) - \frac{R_1 R_F}{R_1 + R_F}(I_{B_2})\right]$$

$$= \frac{R_F^2}{R_1 + R_F}(I_{B_1} - I_{B_2})$$

Since R_F^2 is much larger than $R_1 + R_F$ and since $I_{B_1} - I_{B_2} = I_{io}$, then

$$V_{oo} \cong R_F I_{io} \qquad\qquad (4\text{-}7)$$

where V_{oo} is either positive or negative, depending on whether I_{B_1} is the larger or I_{B_2} is the larger of the two bias currents.

APPENDIX L

SELECTION OF
COUPLING CAPACITORS

The value of coupling capacitance C used between ac amplifier stages is determined by the required low frequency response and the dynamic output and input resistances of the stages being coupled. Generally, lower frequencies and smaller resistances require larger coupling capacitors.

If f_1 is the required low end of the bandwidth (that is, the low frequency at which the signal being coupled is 0.707 of its value at medium frequencies), then

$$C = \frac{1}{2\pi f_1 (R_o + R_i)}$$

where C is the capacitance between the two stages being coupled

R_o is the effective ac output resistance of the first stage, and

R_i is the effective ac input resistance of the second stage.

APPENDIX M

OP AMP APPLICATIONS

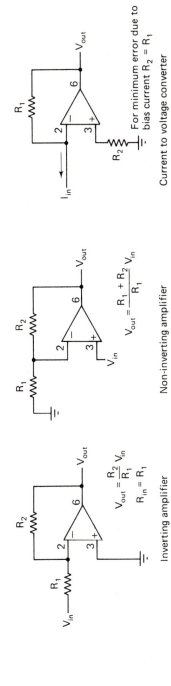

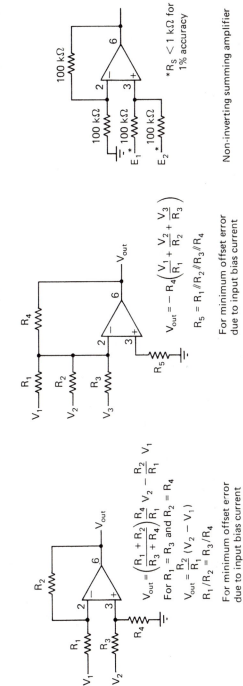

Current to voltage converter

For minimum error due to bias current $R_2 = R_1$

$*R_S < 1\ k\Omega$ for 1% accuracy

Non-inverting summing amplifier

Non-inverting amplifier

$$V_{out} = \frac{R_1 + R_2}{R_1} V_{in}$$

Inverting amplifier

$$V_{out} = \frac{R_2}{R_1} V_{in}$$
$$R_{in} = R_1$$

Inverting summing amplifier

$$V_{out} = -R_4\left(\frac{V_1}{R_1} + \frac{V_2}{R_2} + \frac{V_3}{R_3}\right)$$
$$R_5 = R_1 /\!/ R_2 /\!/ R_3 /\!/ R_4$$

For minimum offset error due to input bias current

Difference amplifier

$$V_{out} = \left(\frac{R_1 + R_2}{R_3 + R_4}\right)\frac{R_4}{R_1} V_2 - \frac{R_2}{R_1} V_1$$
For $R_1 = R_3$ and $R_2 = R_4$
$$V_{out} = \frac{R_2}{R_1}(V_2 - V_1)$$
$$R_1/R_2 = R_3/R_4$$

For minimum offset error due to input bias current

Courtesy of National Semiconductor Corporation

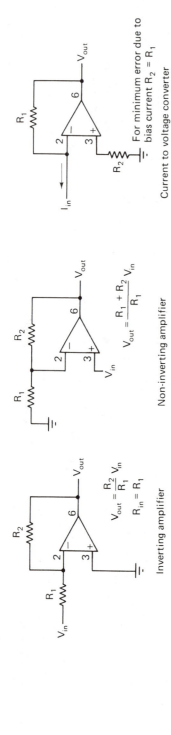

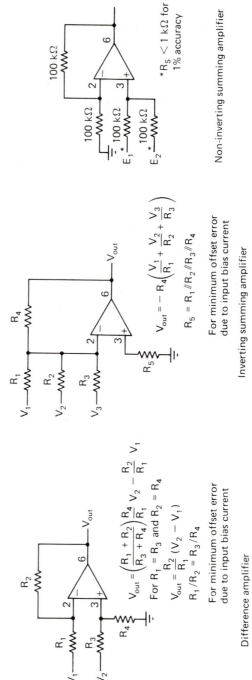

Current to voltage converter

For minimum error due to bias current $R_2 = R_1$

Non-inverting summing amplifier

$^*R_S < 1\ k\Omega$ for 1% accuracy

Non-inverting amplifier

$$V_{out} = \frac{R_1 + R_2}{R_1} V_{in}$$

Inverting summing amplifier

$$V_{out} = -R_4\left(\frac{V_1}{R_1} + \frac{V_2}{R_2} + \frac{V_3}{R_3}\right)$$

$$R_5 = R_1 /\!/ R_2 /\!/ R_3 /\!/ R_4$$

For minimum offset error due to input bias current

Inverting amplifier

$$V_{out} = \frac{R_2}{R_1} V_{in}$$

$$R_{in} = R_1$$

Difference amplifier

$$V_{out} = \left(\frac{R_1 + R_2}{R_3 + R_4}\right)\frac{R_4}{R_1} V_2 - \frac{R_2}{R_1} V_1$$

For $R_1 = R_3$ and $R_2 = R_4$

$$V_{out} = \frac{R_2}{R_1}(V_2 - V_1)$$

$$R_1 / R_2 = R_3 / R_4$$

For minimum offset error due to input bias current

276

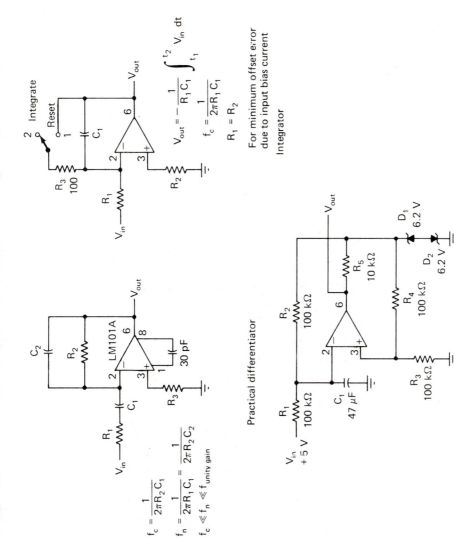

Integrator

$$V_{out} = -\frac{1}{R_1 C_1} \int_{t_1}^{t_2} V_{in}\, dt$$

$$f_c = \frac{1}{2\pi R_1 C_1}$$

$$R_1 = R_2$$

For minimum offset error
due to input bias current

Practical differentiator

$$f_c = \frac{1}{2\pi R_2 C_1}$$

$$f_n = \frac{1}{2\pi R_1 C_1} = \frac{1}{2\pi R_2 C_2}$$

$$f_c \ll f_n \ll f_{unity\ gain}$$

Pulse width modulator

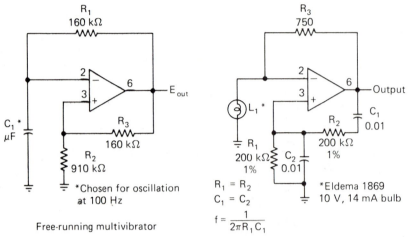

R_1
160 kΩ

2 −
3 +
6 E_{out}

C_1 *
μF

R_3
160 kΩ

R_2
910 kΩ

*Chosen for oscillation
at 100 Hz

Free-running multivibrator

R_3
750

2 −
3 +
6 Output

L_1 *

R_1
200 kΩ
1%

C_2
0.01

R_2
200 kΩ
1%

C_1
0.01

*Eldema 1869
10 V, 14 mA bulb

$R_1 = R_2$
$C_1 = C_2$

$$f = \frac{1}{2\pi R_1 C_1}$$

Wein bridge sine wave oscillator

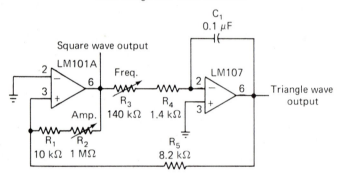

Square wave output

LM101A

2 −
3 +
6

Freq.

R_3
140 kΩ

R_4
1.4 kΩ

C_1
0.1 μF

LM107

2 −
3 +
6

Triangle wave
output

Amp.

R_1
10 kΩ

R_2
1 MΩ

R_5
8.2 kΩ

Function generator

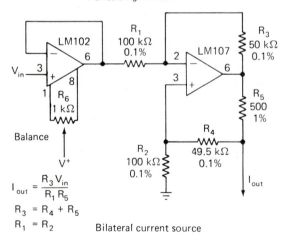

LM102

3 +
6
8
1
R_6
1 kΩ

V_{in}

Balance

V^+

R_1
100 kΩ
0.1%

2 −
3 +
6

LM107

R_3
50 kΩ
0.1%

R_5
500
1%

R_4
49.5 kΩ
0.1%

R_2
100 kΩ
0.1%

I_{out}

$$I_{out} = \frac{R_3 V_{in}}{R_1 R_5}$$

$R_3 = R_4 + R_5$
$R_1 = R_2$

Bilateral current source

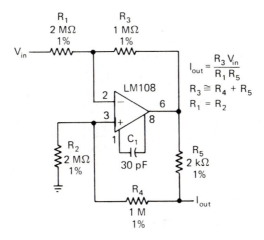

Bilateral current source

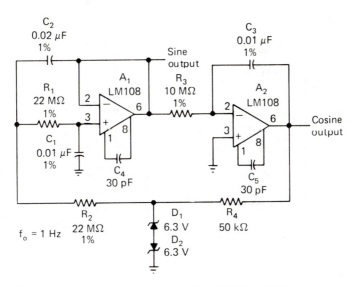

Low frequency sine wave generator with quadrature output

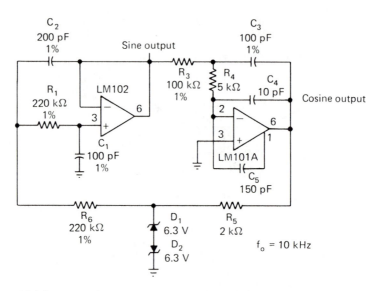

High frequency sine wave generator with quadrature output

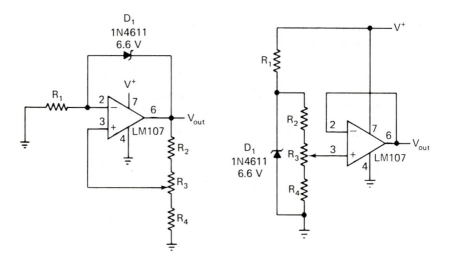

Positive voltage reference   Positive voltage reference

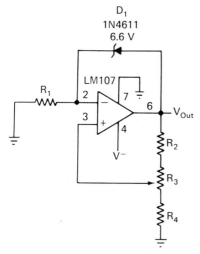

Negative voltage reference

Negative voltage reference

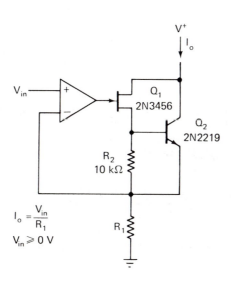

$$I_o = \frac{V_{in}}{R_1}$$

$$V_{in} \geqslant 0 \text{ V}$$

Precision current sink

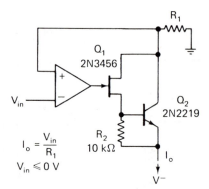

$$I_o = \frac{V_{in}}{R_1}$$
$$V_{in} \leq 0 \text{ V}$$

Precision current source

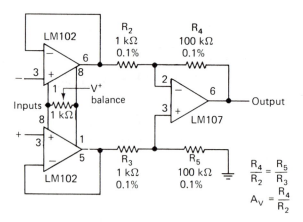

$$\frac{R_4}{R_2} = \frac{R_5}{R_3}$$
$$A_V = \frac{R_4}{R_2}$$

Differential-input instrumentation amplifier

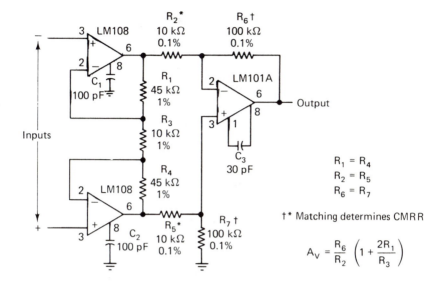

$R_1 = R_4$
$R_2 = R_5$
$R_6 = R_7$

†* Matching determines CMRR

$$A_V = \frac{R_6}{R_2}\left(1 + \frac{2R_1}{R_3}\right)$$

Differential input instrumentation amplifier
with high common mode rejection

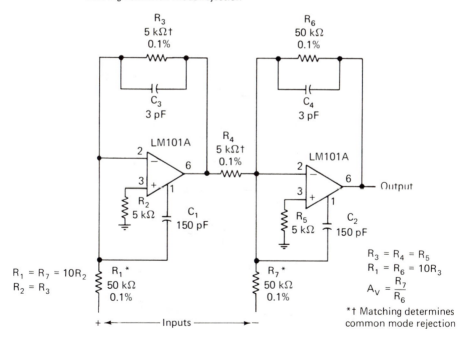

$R_3 = R_4 = R_5$
$R_1 = R_6 = 10R_3$

$$A_V = \frac{R_7}{R_6}$$

$R_1 = R_7 = 10R_2$
$R_2 = R_3$

*† Matching determines
common mode rejection

Instrumentation amplifier with ± 100 volt common mode range

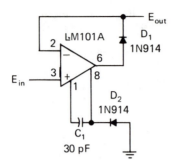

Precision diode

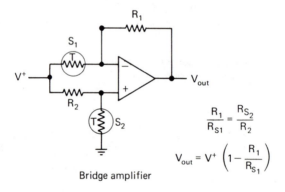

$$\frac{R_1}{R_{S1}} = \frac{R_{S2}}{R_2}$$

$$V_{out} = V^+ \left(1 - \frac{R_1}{R_{S1}}\right)$$

Bridge amplifier

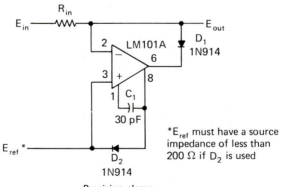

*E_{ref} must have a source impedance of less than 200 Ω if D_2 is used

Precision clamp

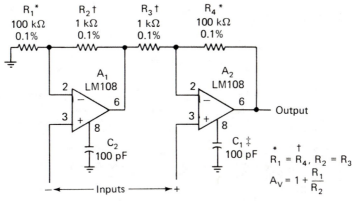

High input impedance instrumentation amplifier

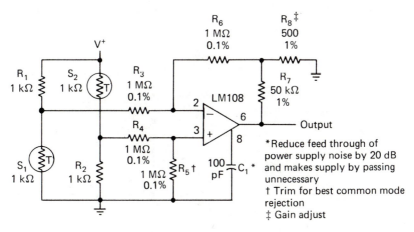

Bridge amplifier with low noise compensation

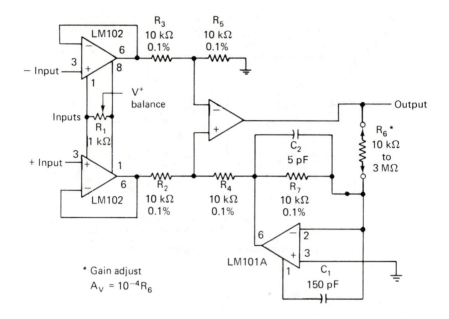

Variable gain, differential-input instrumentation amplifier

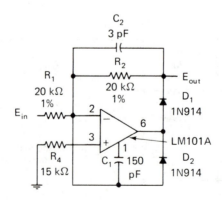

Fast half wave rectifier

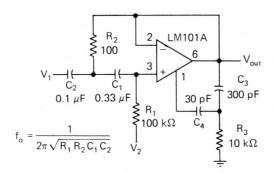

$$f_o = \frac{1}{2\pi \sqrt{R_1 R_2 C_1 C_2}}$$

Tuned circuit

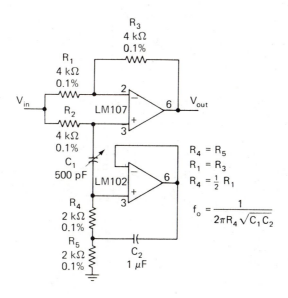

$$R_4 = R_5$$
$$R_1 = R_3$$
$$R_4 = \tfrac{1}{2} R_1$$

$$f_o = \frac{1}{2\pi R_4 \sqrt{C_1 C_2}}$$

Easily tuned notch filter

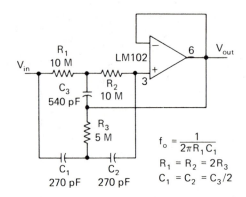

High Q notch filter

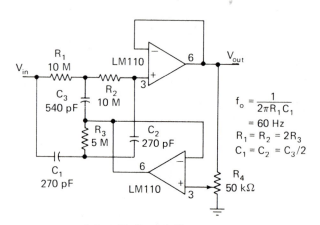

Adjustable Q notch filter

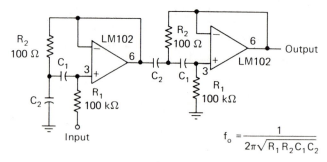

Two-stage tuned circuit

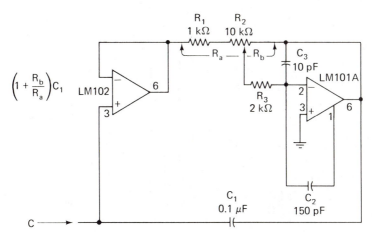

Variable capacitance multiplier

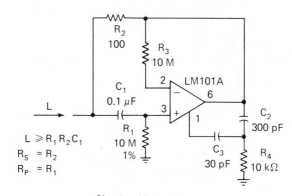

Simulated inductor

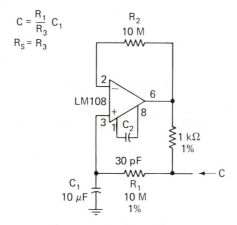

Capacitance multiplier

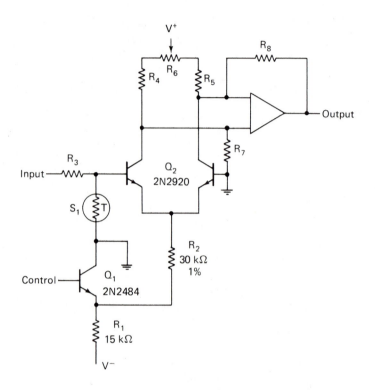

Voltage controlled gain circuit

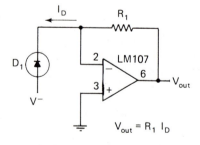

$$V_{out} = R_1 I_D$$

Photodiode amplifier

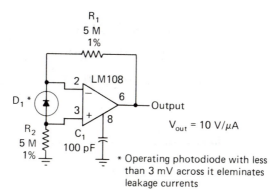

Photodiode amplifier

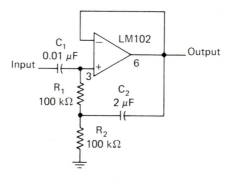

High input impedance ac follower

*APPLICATIONS
OF LM 2900/3900
QUAD AMPLIFIERS*

COLLECTION OF CURRENT-DIFFERENCING (NORTON) AMPLIFIER CIRCUITS

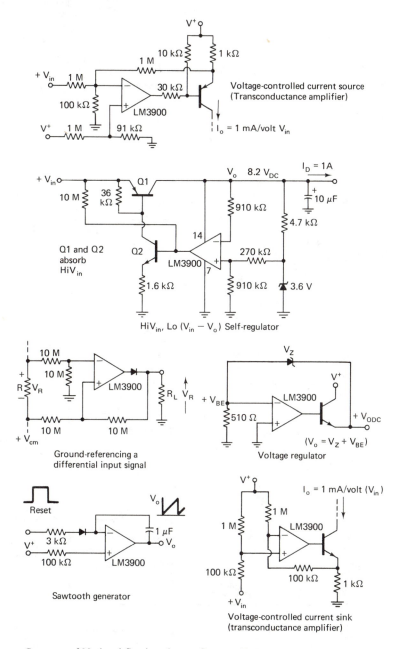

Courtesy of National Semiconductor Corporation

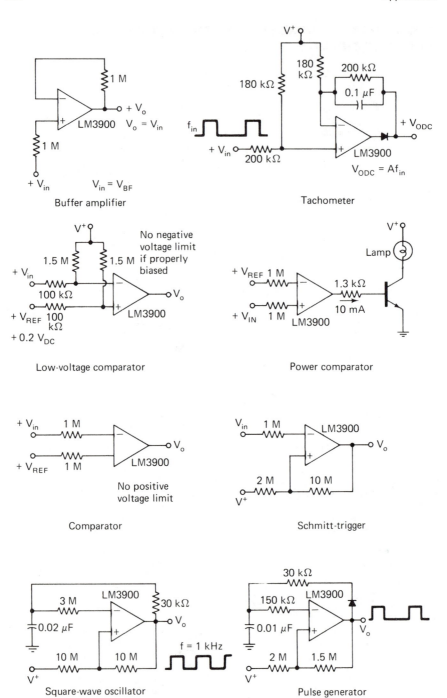

Buffer amplifier

Tachometer

Low-voltage comparator

Power comparator

Comparator

Schmitt-trigger

Square-wave oscillator

Pulse generator

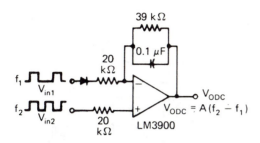

Frequency differencing tachometer

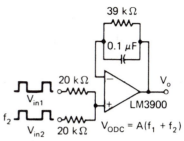

Frequency averaging tachometer

$V_{ODC} = A(f_2 - f_1)$

$V_{ODC} = A(f_1 + f_2)$

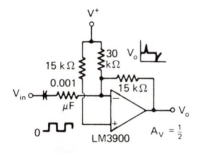

Squaring amplifier (W/hysteresis)

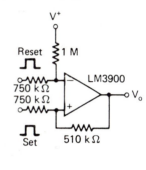

Bi-stable multivibrator

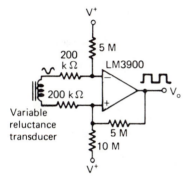

Differentiator (common-mode biasing keeps input at $+ V_{BE}$)

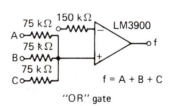

"OR" gate

$f = A + B + C$

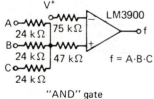

"AND" gate

$f = A \cdot B \cdot C$

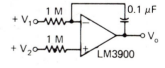

Difference integrator

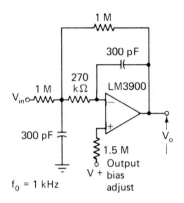

Low-pass active filter

Staircase generator

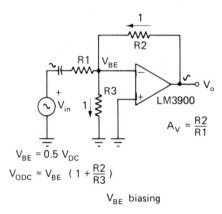

$$A_v = \frac{R2}{R1}$$

$$V_{BE} = 0.5\, V_{DC}$$

$$V_{ODC} = V_{BE}\left(1 + \frac{R2}{R3}\right)$$

V_{BE} biasing

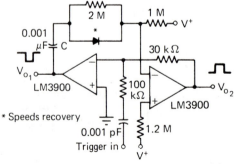

One-shot multivibrator

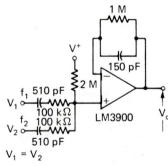

Low-frequency mixer

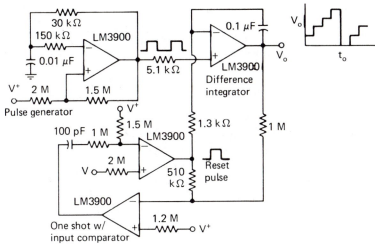

Free-running staircase generator/pulse counter

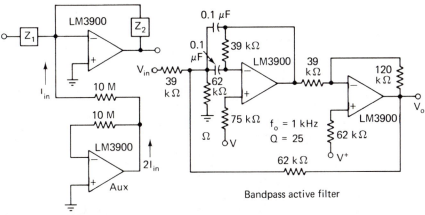

Supplying I_{in} with aux. amp
(to allow high Z feedback networks)

Bandpass active filter

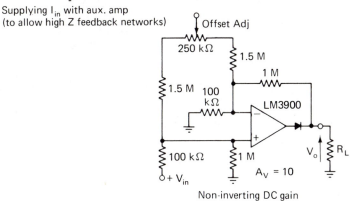

Non-inverting DC gain

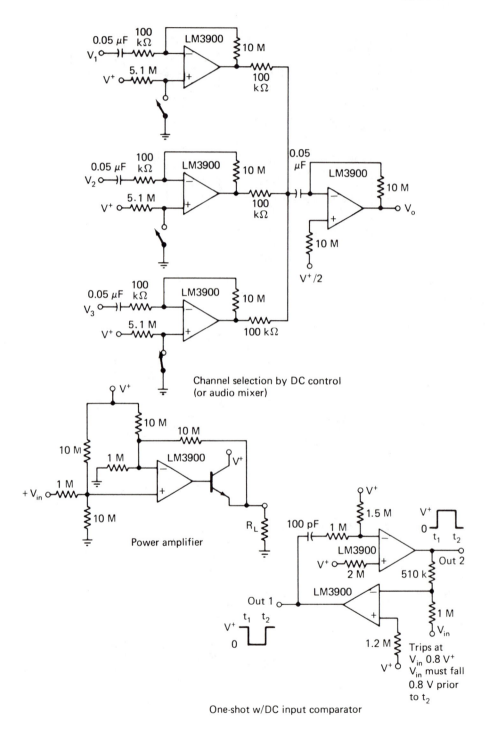

Channel selection by DC control
(or audio mixer)

Power amplifier

One-shot w/DC input comparator

APPENDIX O

Program in BASIC that prints or displays tables of gain vs. frequency characteristics for single Op Amp second-order low pass and high pass active filters.

```
0010 PRINT "PLOT OF A 2ND ORDER LOW PASS ACTIVE FILTER"
0020 PRINT
0030 PRINT "WHAT IS THE VALUE OF R1 IN OHMS?"
0040 INPUT R1
0050 PRINT "WHAT IS THE VALUE OF R2 IN OHMS?"
0060 INPUT R2
0070 PRINT "WHAT IS THE VALUE OF C1 IN FARADS?"
0080 INPUT C1
0090 PRINT "WHAT IS THE VALUE OF C2 IN FARADS?"
0100 INPUT C2
0110 PRINT "WHAT IS THE CLOSED-LOOP GAIN OF THE OP AMP?"
0120 INPUT B
0130 PRINT "WHAT IS THE INITIAL FREQUENCY?"
0140 INPUT F
0150 PRINT "BY WHAT MULTIPLES DO YOU WANT TO INCREASE F?"
0160 INPUT M
0170 PRINT
0180 LET D=1/(R1*R2*C1*C2)^.5
0190 LET A=(R2*C2+R1*C2+R1*C1*(1-B))*D
0200 PRINT "THE DAMPING COEFFICIENT D=";A
0210 PRINT
0220 PRINT "FREQ","GAIN","GAIN","PHASE"
0230 PRINT   TAB(2);"HZ","OUT/IN"," DB"," DEG"
0240 PRINT "_____"
0250 LET W=2*3.14159*F*(R1*R2*C1*C2)^.5
0260 LET K=ABS(1-W^2)
0270 LET G=B/(K^2+(A*W)^2)^.5
0280 LET G1=20/2.30259*LOG(G)
0290 LET O=-ATN(A*W/K)*57.2958
0300 PRINT F,G,G1,O
0310 LET F=F*M
0320 GOTO 0250
0330 END
*
```

```
0010 PRINT "PLOT OF A 2ND ORDER HIGH PASS ACTIVE FILTER"
0020 PRINT
0030 PRINT "WHAT IS THE VALUE OF R1 IN OHMS?"
0040 INPUT R1
0050 PRINT "WHAT IS THE VALUE OF R2 IN OHMS?"
0060 INPUT R2
0070 PRINT "WHAT IS THE VALUE OF C1 IN FARADS?"
0080 INPUT C1
0090 PRINT "WHAT IS THE VALUE OF C2 IN FARADS?"
0100 INPUT C2
0110 PRINT "WHAT IS THE CLOSED-LOOP GAIN OF THE OP AMP?"
0120 INPUT B
0130 PRINT "WHAT IS YOUR INITIAL MAXIMUM FREQUENCY?"
0140 INPUT F
0150 PRINT "BY WHAT MULTIPLES DO YOU WANT TO DECREASE THE FREQUENCY?"
0160 INPUT M
0170 PRINT
0180 LET D=1/(R1*R2*C1*C2)^.5
0190 LET A=(R1*C1+R1*C2+R2*C2*(1-B))*D
0200 PRINT "THE DAMPING COEFFICIENT D=";A
0210 PRINT
0220 PRINT "FREQ","GAIN","GAIN","PHASE"
0230 PRINT  TAB(2);"HZ","OUT/IN"," DB"," DEG"
0240 PRINT "_____"
0250 LET W=2*3.14159*F*(R1*R2*C1*C2)^.5
0260 LET K=ABS(1-W^2)
0270 LET G=B*W^2/(K^2+(A*W)^2)^.5
0280 LET G1=20/2.30259*LOG(G)
0290 LET O=-ATN(A*W/K)*57.2958
0300 PRINT F,G,G1,O
0310 LET F=F/M
0320 GOTO 0250
0330 END
*
```

APPENDIX P

Program in BASIC that prints or displays a table of gain vs. frequency characteristics of a multiple-feedback bandpass active filter.

```
0010 PRINT "GIVE ME YOUR COMPONENT VALUES FOR A MULTIPLE FEEDBACK"
0020 PRINT "BANDPASS ACTIVE FILTER."
0030 PRINT
0040 PRINT "WHAT IS THE VALUE OF R1 IN OHMS?"
0050 INPUT R1
0060 PRINT "WHAT IS THE VALUE OF R2 IN OHMS?"
0070 INPUT R2
0080 PRINT "WHAT IS THE VALUE OF THE FEEDBACK RESISTOR RF?"
0090 INPUT R
0100 PRINT "MAKE THE CAPACITORS EQUAL. WHAT IS THE VALUE OF EACH CAPACITOR?"
0110 INPUT C
0120 PRINT "WHAT IS THE INITIAL FREQUENCY F?"
0130 INPUT F
0140 PRINT "BY WHAT INCREMENTS DO YOU WANT TO INCREASE THE FREQUENCY?"
0150 INPUT B
0160 LET W=((R1+R2)/(R1*R2*R*C^2))^.5
0170 LET W1=W/(2*3.14159)
0180 LET P=2/(R*C*2*3.14159)
0190 LET Q=W1/P
0200 LET S=(1/(4*Q^2)+1)^.5
0210 LET S1=1/(2*Q)
0220 LET A=-R/(2*R1)
0230 LET D=1/Q
0240 LET F1=W1*(S-S1)
0250 LET F2=W1*(S+S1)
0260 PRINT "THE CENTER FREQUENCY IS";W1;"HZ"
0270 PRINT
0280 PRINT "THE BANDWIDTH IS";P;"HZ"
0290 PRINT
0300 PRINT "THE Q IS";Q
0310 PRINT
0320 PRINT "F1=";F1;"HZ AND F2=";F2;"HZ"
0330 PRINT
0340 PRINT "FREQ","GAIN","GAIN(DB)"
0350 PRINT "------------------------------------------"
0360 PRINT
0370 LET W1=2*3.14159*F
0380 LET K=(A*D*W*W1)^2
0390 LET K1=W1^4+W^2*(D^2-2)*W1^2+W^4
0400 LET G=(K/K1)^.5
0410 LET G1=20/2.30259*LOG(G)
0420 PRINT F,G,G1
0430 LET F=F+B
0440 GOTO 0370
0450 END
```

Program in BASIC that prints or displays a table of gain vs. frequency characteristics of a multiple-feedback bandstop (notch) filter.

```
0010 PRINT "GIVE ME YOUR COMPONENT VALUES FOR A MULTIPLE FEEDBACK"
0020 PRINT "NOTCH FILTER"
0030 PRINT
0040 PRINT "WHAT IS THE VALUE OF R1 IN OHMS?"
0050 INPUT R1
0060 PRINT "WHAT IS THE VALUE OF R2 IN OHMS?"
0070 INPUT R2
0080 PRINT "WHAT IS THE VALUE OF THE FEEDBACK RESISTOR RF?"
0090 INPUT R
0100 PRINT "WHAT IS THE VALUE OF R5?"
0110 INPUT R5
0120 PRINT "WHAT IS THE VAUE OF R4?"
0130 INPUT R4
0140 PRINT "WHAT IS THE VALUE OF R3?"
0150 INPUT R3
0160 PRINT "MAKE THE CAPACITORS EQUAL. WHAT IS THE VALUE OF EACH CAPACITOR?"
0170 INPUT C
0180 PRINT "WHAT IS THE INITIAL FREQUENCY F?"
0190 INPUT F
0200 PRINT "BY WHAT INCREMENTS DO YOU WANT TO INCREASE THE FREQUENCY?"
0210 INPUT B
0220 LET W=((R1+R2)/(R1*R2*R*C^2))^.5
0230 LET W1=W/(2*3.14159)
0240 LET P=2/(R*C*2*3.14159)
0250 LET Q=W1/P
0260 LET S=(1/(4*Q^2)+1)^.5
0270 LET S1=1/(2*Q)
0280 LET A=-R/(2*R1)
0290 LET D=1/Q
0300 PRINT "THE CENTER FREQUENCY IS";W1;"HZ"
0310 PRINT
0320 PRINT
0330 PRINT "FREQ","GAIN","GAIN(DB)"
0340 PRINT "_____"
0350 PRINT
0360 LET W1=2*3.14159*F
0370 LET K=(A*D*W*W1)^2
0380 LET K1=W1^4+W^2*(D^2-2)*W1^2+W^4
0390 LET G=(K/K1)^.5
0400 LET G1=R5/R3-G*R5/R4
0410 LET G1=ABS(G1)
0420 LET G2=20/2.30259*LOG(G1)
0430 PRINT F,G1,G2
0440 LET F=F+B
0450 GOTO 0360
0460 END
```

APPENDIX R

DERIVATION OF EQUATION (6-1)

Figure 2-4 of Chapter 2 shows that an Op Amp is equivalent to a signal source $(A_{VOL})V_{id}$ in series with a resistance (R_o), as the load (R_L) "sees" it. An Op Amp also has internal capacitances. Their effect on the Op Amp's output characteristics is the same as the effect of a single *equivalent* capacitance C that is across the output and ground (see Fig. R-1). At low frequencies, the reactance of this capacitance C is so large that it acts like an open and has no influence on the amplitude of the output signal voltage V_o. At higher frequencies, however, the reactance of C decreases, causing the amplitude of the output V_o to decrease. A typical V_o vs. frequency curve is shown in Fig. R-2. Such curves can be plotted by measuring V_o with various frequencies of input signals V_s, while the amplitude of the input signal V_s is kept constant. Note that f_c is the frequency at which V_o decreases to 0.707 of its maximum value. This is a 3-dB decrease. Frequency f_c is called the *critical frequency, cutoff frequency, half-power frequency,* or *3-dB-down frequency*, to name just a few of the more common terms. As shown in Fig. R-2, f_c is the upper limit of an amplifier's bandwidth.

In the equivalent circuit of Fig. R-1, the reactance of C is equal to the resistance of R at the frequency f_c. Since, generally,

$$X_c = \frac{1}{2 \pi f C}$$

then

$$X_c = R = \frac{1}{2 \pi f_c C}$$

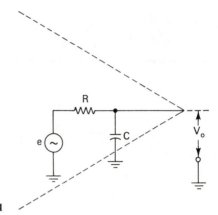

Figure R-1

and therefore,

$$f_c = \frac{1}{2 \pi RC} \qquad (9\text{-}6)$$

Since Op Amps can amplify down to 0 Hz and f_c marks the upper limit of the bandwidth, we can show that the bandwidth $BW = f_c$. The above equation can thus be modified to

$$BW = \frac{1}{2\pi RC} \qquad (R\text{-}1)$$

If a relatively high frequency squarewave is applied to the input of an amplifier, the equivalent circuit and resulting output are as shown in Fig. R-3a and b. Note that V_o rises to $0.1\ V_{max}$ at time t_1 and that V_o reaches $0.9\ V_{max}$ at t_2. Therefore, the rise time

$$T_R = t_2 - t_1 \qquad (R\text{-}2)$$

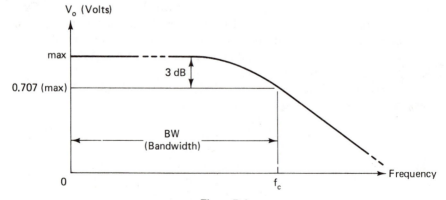

Figure R-2

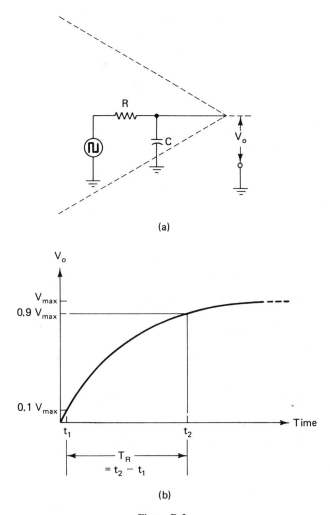

(a)

(b)

Figure R-3

Since this output rises exponentially, it can generally be expressed with the
equation

$$V_{o(t)} = V_{max}(1 - e^{-t/RC})$$

Specifically at time t_1, then,

$$0.1\ V_{max} = V_{max}(1 - e^{-t_1/RC}) \tag{R-3}$$

Dividing both sides of Eq. (R-3) by V_{max} yields

$$0.9 = e^{-t_1/RC}$$

Finding the natural logarithm of both sides shows that

$$0.1 \cong t_1/RC$$

and, therefore,

$$t_1 \cong 0.1\ RC \qquad\qquad (R\text{-}4)$$

Similarly, at t_2,

$$0.9\ V_{\max} = V_{\max}(1 - e^{-t_2/RC})$$

Dividing by $V_{\max}$ and rearranging yields

$$0.1 = e^{-t_2/RC}$$

and, thus

$$2.3 \cong t_2/RC$$

and

$$t_2 \cong 2.3\ RC \qquad\qquad (R\text{-}5)$$

By substituting Eqs. (R-4) and (R-5) into Eq. (R-2) we can show that

$$T_R \cong 2.3\ RC - 0.1\ RC = 2.2\ RC$$

and thus

$$RC \cong T_R/2.2$$

This last equation substituted into Eq. (R-1) gives us

$$BW \cong \frac{1}{2\ \pi\ T_R/2.2}$$

or

$$BW \cong \frac{0.35}{T_R} \qquad\qquad (6\text{-}1)$$

Similar algebraic steps will show that also $BW \cong 0.35/T_F$.

PHASE SHIFT IN SECOND-ORDER
ACTIVE FILTERS

The amplitude of the output V_{out} vs. frequency is usually most important in active filter applications. Where phase shift ϕ between V_{in} and V_{out} is also of concern, the phase angle of V_{out} with respect to V_{in} is

$$\phi = -\arctan\left[\frac{2\,\pi\,f(R_1 C_2 + R_2 C_2 + R_1 C_1(1 - A_v))}{|1 - 2\,\pi\,f\sqrt{R_1 R_2 C_1 C_2}|}\right]$$

with the LP filter types of Fig. 9-9 and 9-15, and where

$$A_v = \frac{R_f}{R_i} + 1$$

With HP filter types of Figs. 9-12 and 9-16, the phase shift

$$\phi = -\arctan\left[\frac{2\,\pi\,f\,(R_1 C_1 + R_1 C_2 + R_2 C_2(1 - A_v))}{|1 - 2\,\pi\,f\sqrt{R_1 R_2 C_1 C_2}|}\right]$$

The BASIC programs of Appendix O include solutions of phase shift ϕ.

ANSWERS TO SELECTED
ODD-NUMBERED PROBLEMS

Chapter 1. *1-1.* $I_B \cong 6.25 \ \mu A$. *1-3.* $V_{C_1} \cong V_{C_2} \cong 6.4$ V. *1-5.* $72. \le V_{od}/V_{id} \le 144$.

Chapter 2. *2-1.* If a 115-V/15-V transformer of Table 2-2 is used, $R_1 = R_2 \cong 93.6 \ \Omega$; use standard 82-$\Omega$ resistors with 1 W or greater rating. Use 1 W or larger 10-V zeners. $C_1 = C_2 \cong 570 \ \mu F$ or larger. Rectifier diodes with PIV of 50 V or more. *2-3.* No, because the load draws more than the secondary current rating.

Chapter 3. *3-1.* (b). *3-3.* (e). *3-5.* $A_v = 2$, V_o has waveform (a). *3-7.* $A_v = 101$, V_o has waveform (f). *3-9.* As seen looking to the right of point x, the circuit of Fig. 3-17a has the more constant input resistance. *3-11.* About 20 mV dc. *3-13.* About 3 V dc. *3-15.* $R_{o(eff)} = 0.375 \ \Omega$. *3-17.* $V_o = 400$ mV dc. *3-19.* $A_{v(min)} = 2$, $A_{v(max)} = 202$. *3-21.* Waveform V_o is unclipped, 180° out of phase with the input but with peaks 150 times larger. *3-23.* Waveform V_o is a sawtooth for the beginning alternations and is 180° out of phase with the input. It is clipped at plus or minus 5.5 V at times t_4, t_5, t_6, etc. *3-25.* The output is in phase with the input and is not clipped on the first alternation that has a peak twice that of the input. The remaining alternations are clipped at ± 3 V. *3-27.* The output is out of phase with the input and is not clipped on the first two alternations. The remaining alternations are clipped at about ± 13 V.

Chapter 4.   *4-1.* −40.   *4-3.* −101.   *4-5.* Between 200 and 250 mV dc.   *4-7.* About 20 mV dc.   *4-9.* Between 220 mV and 280 mV dc.   *4-11.* 990 Ω or about 1 kΩ.   *4-13.* About 20 mV dc.   4-15. Replace the 200-kΩ resistor with a smaller one or replace the 100-Ω resistor with a larger resistance value.   *4-17.* 0.2 V.   *4-19.* 0.3 V.   *4-21.* Provides capability to reduce (null) the output to 0 V.   *4-23.* About 6.2 V dc.

Chapter 5.   *5-1.* 80 dB.   *5-3.* (a) Output varies from 3 V to −2 V, (b) Output noise is 1 V rms-60 Hz because there is no *CMRR* with this circuit since it is not wired in differential mode.   *5-5.* (a) 1.82 V, (b) −5 V to 5 V, (c) 20 μV rms @ 60 Hz.   *5-7.* (a) 9 V, (b) 8.18 V.   *5-9.* (a) 0 V dc, (b) −1.58 V to 1.43 V.   *5-11.* 15 V to −15 V.   *5-13.* About 7.6/1.   *5-15.* −11.   *5-17.* V_{CM} = 5.3 V, V_o = -4.4 V dc.   *5-19.* 600 to 5700.   *5-21.* 100.   *5-23.* (a) A_v = A_{VOL}, (b) A_v = 4.4.   *5-25.* R_a provides gain adjustment, R_4 provides NULL adjustment.   *5-27.* About 3.48 V.

Chapter 6.   *6-1.* 1 MHz.   *6-3.* C_1 = 500 pF, R_1 = 1.5 kΩ, C_2 = 20 pF.   *6-5.* 1 MHz.   *6-7.* About 50 kHz.   *6-9.* A_v = A_{VOL}, BW is insignificant.   *6-11.* −20 dB/decade or −6 dB/octave.   *6-13.* About 20 kHz.   *6-15.* 0.2 μV rms @ 60 Hz.   *6-17.* V_o is a sawtooth with 4-V peak-to-peak amplitude.   *6-19.* About 117 kHz.   *6-21.* V_{no} ≅ 50 μV rms.   *6-23.* 40 kHz.   *6-25.* About 60 kHz.   *6-27.*   (See below.)

```
0010 PRINT
0020 PRINT "PLOT OF A WAVEFORM THAT RESULTS FROM THE SUM OF"
0030 PRINT "A SINEWAVEAND SEVERAL OF ITS HARMONICS"
0040 FOR X=.1 TO 12.6 STEP .1
0050    LET V1=-1*SIN(X)
0070    LET V3=-.333333*SIN(3*X)
0090    LET V5=-.2*SIN(5*X)
0110    LET K=15
0120    LET V7=3
0130    LET V=K*(V1+V2+V3+V4+V5+V6+V7)
0140    PRINT TAB(V)"@"
0150 NEXT X
0160 END
```

Chapter 7.   *7-1.* ΔV_{io} ≅ ±40 μV, ΔV_{oo} ≅ ±8 mV.   *7-3.* About 8 μV rms.   *7-5.* 10^{-5}.   *7.7.* ΔV_{oo} ≅ 4.8 mV.   *7-9.* The output drift from 500 mV is by about 0.2 mV in either case.   *7-11.* (a) About 0.651 mV, (b) About 0.656 mV.   *7-13.* (a) About 97.5ns, (b) About 110 ns.   *7-15.* (a) About 200 mV for an LM741C or 48 mV for an LM741E, (b) About 68 μV.

Chapter 8.   *8-1.* $A_v = 21$.   *8-3.* $A_v = 11$.   *8-5.* $A_v = -1$.   *8-7.* (a) 9.4 V, (b) 5.55 V.   *8-9.* 0.4 V.   *8-11.* About 22 V to 5.45 V.   *8-13.* $R_F/R_1 = 0.7$, $R_a/R_b < 0.8$, $R_s \cong 22\ \Omega$, $P_{C(max)} \cong 21$ W, max $P_z = 375$ mW.   *8-15.* -2.5 V, 570 Ω.   *8-17.* 20 s.

Chapter 9.   *9-1.* $f_c \cong 603$ Hz, $R_F = R = 1.2$ kΩ.   *9-3.* (a) 0.453 dB, (b) -1.58 dB, (c) -2.99 dB, (d) -4.41 dB, (e) -5.74 dB.   *9-5.* 0 dB to 600 Hz, after 600 Hz, roll-off is at a -20 dB/decade rate.   *9-7.* $d = 0.899$, LP Chebyshev.   *9-9.* $f_x = f_p = 1505$ Hz.   *9-11.* $C_2 = 0.1\ \mu$F, $R_1 = R_2$ 1406.74 Ω.   *9-13.* $d = 0.77$, 3-dB Chebyshev HP filter.   *9-15.* $f_p = 251.9$ Hz.   *9-17.* $R_1 = 7957.7\ \Omega$, $R_2 = 418.8\ \Omega$, $R_F = 159.155$ kΩ, BW = 100 Hz, $f_1 = 951.28$ Hz, $f_2 = 1051.28$ Hz.   *9-19.* 1031 Hz.   *9-21.* $R_1 = 39.78$ kΩ, $R_2 = 812\ \Omega$, $R_F = 79.58$ kΩ, $R_3 = 1$ kΩ, $R_4 = 1$ kΩ and $R_5 = 10$ kΩ.

Chapter 10.   *10-1.* Without the zeners, the output would be a sinewave with 10-V peaks. With these zeners, both the positive and negative alternations are clipped at about 6 V.   *10-3.* Without the diodes, the output would be sinusoidal peaking at 11 V. With the diodes, positive alternations are clipped at about 4 V, but negative alternations are unclipped.   *10-5.* Zener D_1 is forward biased when the output attempts to swing negatively causing $V_o = 0$ V at such times. The positive alternations are sinusoidal and are clipped at about 4 V.   *10-7.* 2000/1.   *10.9.* (a) $V_{o1} = -15$ V, $V_{o2} = 15$ V, $V_{o3} = 15$ V, (b) $V_{o1} = -15$ V, $V_{o2} = -15$ V, $V_{o3} = 15$ V, (c) $V_{o1} = V_{o2} = V_{o3} = -15$ V.   *10-11.* Waveform V_o is a squarewave with the same frequency as the input sinewave. Waveform V_L is a differentiation of waveform V_o; i.e., V_L spikes positively when V_o rises positively, V_L spikes negatively when V_o changes negatively.   *10-13.* Like Fig. 10-10b where $V_a' = -0.429$ V and $V_a = 0.429$. The noise is 0.857 V.   *10-15.* A squarewave with no noise.   *10-17.* The output will look like a full wave rectified waveform having a peak of 3 V.   *10-19.* (a) 1.2 V, (b) -2.8 V.   *10-21.* Output is rectangular with 4.74 ms positive alternations and 5.26 ms negative alternations.

Chapter 11.   *11-1.* 250 Hz.   *11-3.* 26 V.   *11-5.* The sawtooth output V_o' will have its positive and negative alternations clipped.   *11-7.* V_o' will be a nonsymmetrical sawtooth with half of its alternations having a time period 20 times larger than the remaining alternations.   *11-9.* About 1.6 kΩ.

Chapter 12.   *12-1.* $R_1 = 3$ kΩ, $R_2 = 300$ kΩ, $R_F = 150$ kΩ.   *12-3.* $V_o \cong$ 6.67 V, $A_v \cong 45.45$.   *12-5.* $A_v = 45.45$, $V_o = -5.33$ V.   *12-7.* $R_1 = 12$ kΩ, $R_2 = 2.4$ MΩ, $R_F = 1.2$ MΩ.   *12-9.* $A_v = -100$, $V_o = 10$ V, $I_2 = 21.27$ μA.   *12-11.* (a) unlighted, (b) unlighted, (c) unlighted, (d) lighted, (e) lighted, (f) unlighted, (g) unlighted, $V_1 = 3$ V, $V_2 = 9$ V.   *12-13.* 7.32 V, 2.04 V, N.M. $= 5.28$ V.

INDEX